FRANZ A. RŒDELBERGER – VERA I. GROSCHOFF

and 100 wildlife photographers

present

The Wonders of Wildlife in Europe

Nature Observed in 280 Pictures

ENGLISH VERSION BY MARY PHILLIPS AND PETER WHITEHEAD

WITH AN INTRODUCTION BY PETER SCOTT

CONSTABLE · LONDON

VIKING PRESS INC.
NEW YORK

LONGMANS GREEN
TORONTO

Original title: TIERWELT EUROPAS / FÉERIE ANIMALE

Published in 1963 by Constable & Co. Ltd. 10 Orange Street, London WC 2

PRINTED IN SWITZERLAND

INTRODUCTION

by PETER SCOTT

(Chairman of the World Wildlife Fund)

"The extermination of a species of animals is like the destruction of a unique work of art. Just imagine a group of vandals going round the world every year and solemnly taking down all the works of two or three great masters and carefully and deliberately destroying them."

H. R. H. Prince Philip, Duke of Edinburgh, speaking at the World Wildlife Fund Dinner in London on 6th November 1962.

One of the most terrifying attributes of human beings is their relentless capacity to destroy. When man was a creature of the wilderness he had to kill wild animals to survive. The dark primitive forests which were the home of our ancestors may seem a far cry from the suburban life which so many of us in this country endure today; but not much of the violence and savagery of those forests has disappeared from human nature: on the other hand man has gradually evolved a conscience. The constant battle between instinct and conscience is responsible for most of the dilemmas now facing the human race.

Centuries of civilisation and "progress" have entailed the wholesale destruction of wild creatures and their natural habitats. Nearly 100 species of vertebrate animals have disappeared completely from the face of the earth during this century at the hands of man. Now it seems that more and more people are beginning to think this is rather a pity — that maybe we should call a halt to this process of extermination.

In 1961 The World Wildlife Fund was set up to provide an emergency conservation service for threatened animals (and plants) anywhere in the world — a sort of new Noah's Ark; and also, in the longer term, to foster the idea that man himself is an integral part of his natural surroundings, and that he has responsibilities of trusteeship for the natural world over which he now exercises such sweeping power. Emergency action is an urgent need in many cases; but the dissemination of knowledge is just as important a part of any effective, planned policy of conservation; and conservation is not just something that concerns Africa and other far-away places. The problem is on our own doorstep too. We have our rarities in Britain — the Kite and the Osprey, the Pine Martin and the Polecat, the Swallow Tail Butterfly and the Large Blue. In Europe there is the Musk Ox, the Brown Bear, the Spanish Lynx and the Spanish Imperial Eagle.

Conservation is not just just the protection of cuddly animals by sentimentalists. It has a serious ethical side, and it has aesthetic and economic aspects too. It is a scientific principle which concerns every one of us. That is why I particularly welcome this book, THE WONDERS OF WILDLIFE IN EUROPE. It is a magnificent demonstration of the infinite variety of nature which *must* be handed on unimpaired as a continuing source of wonder and delight to mankind far into the future.

PETER SCOTT
Slimbridge,
June 1963.

➢ *Gannet colony, Garden Rocks in Trois Vaux Bay, Alderney, off the coast of Brittany.*

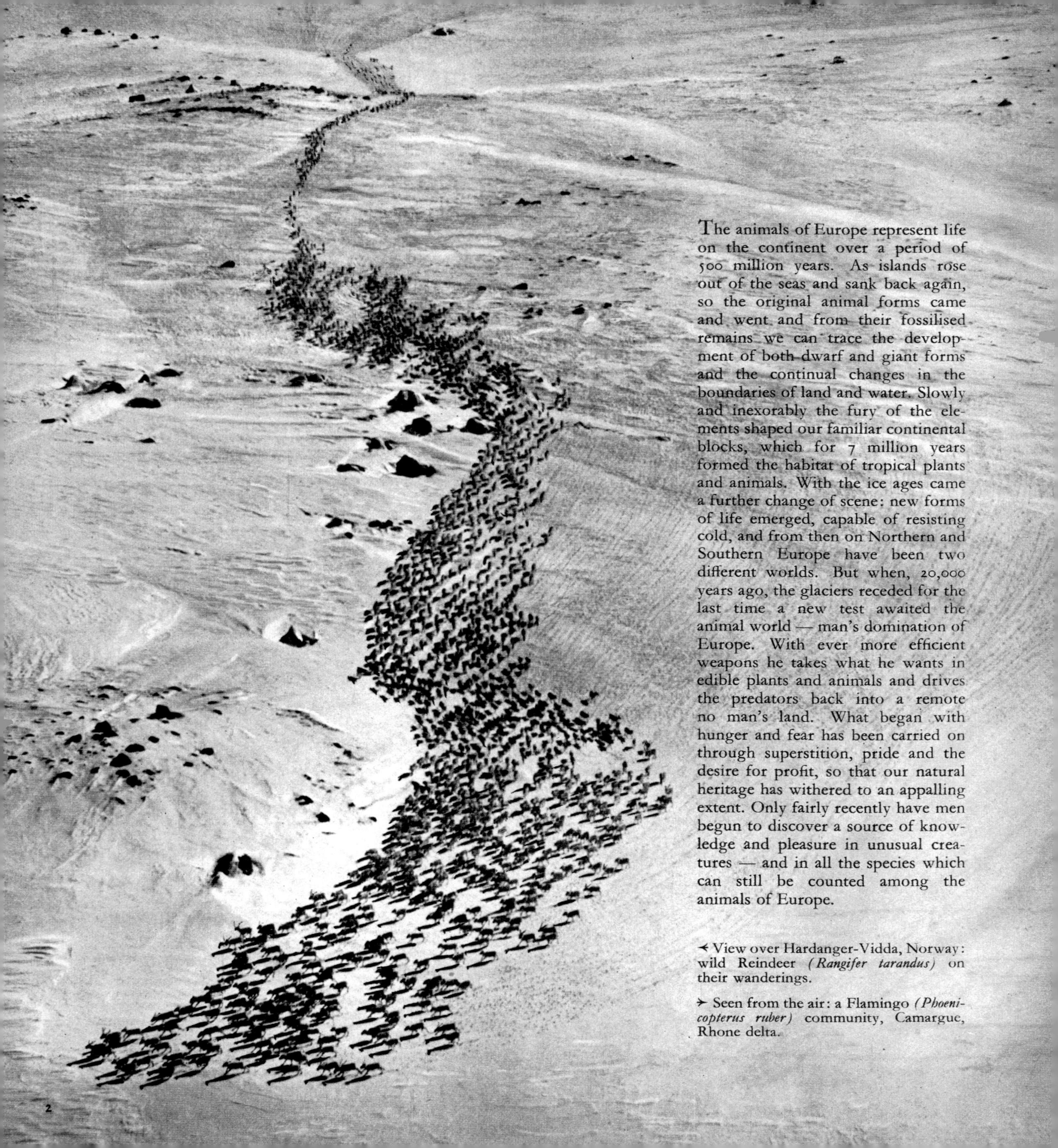

The animals of Europe represent life on the continent over a period of 500 million years. As islands rose out of the seas and sank back again, so the original animal forms came and went and from their fossilised remains we can trace the development of both dwarf and giant forms and the continual changes in the boundaries of land and water. Slowly and inexorably the fury of the elements shaped our familiar continental blocks, which for 7 million years formed the habitat of tropical plants and animals. With the ice ages came a further change of scene: new forms of life emerged, capable of resisting cold, and from then on Northern and Southern Europe have been two different worlds. But when, 20,000 years ago, the glaciers receded for the last time a new test awaited the animal world — man's domination of Europe. With ever more efficient weapons he takes what he wants in edible plants and animals and drives the predators back into a remote no man's land. What began with hunger and fear has been carried on through superstition, pride and the desire for profit, so that our natural heritage has withered to an appalling extent. Only fairly recently have men begun to discover a source of knowledge and pleasure in unusual creatures — and in all the species which can still be counted among the animals of Europe.

← View over Hardanger-Vidda, Norway: wild Reindeer *(Rangifer tarandus)* on their wanderings.

→ Seen from the air: a Flamingo *(Phoenicopterus ruber)* community, Camargue, Rhone delta.

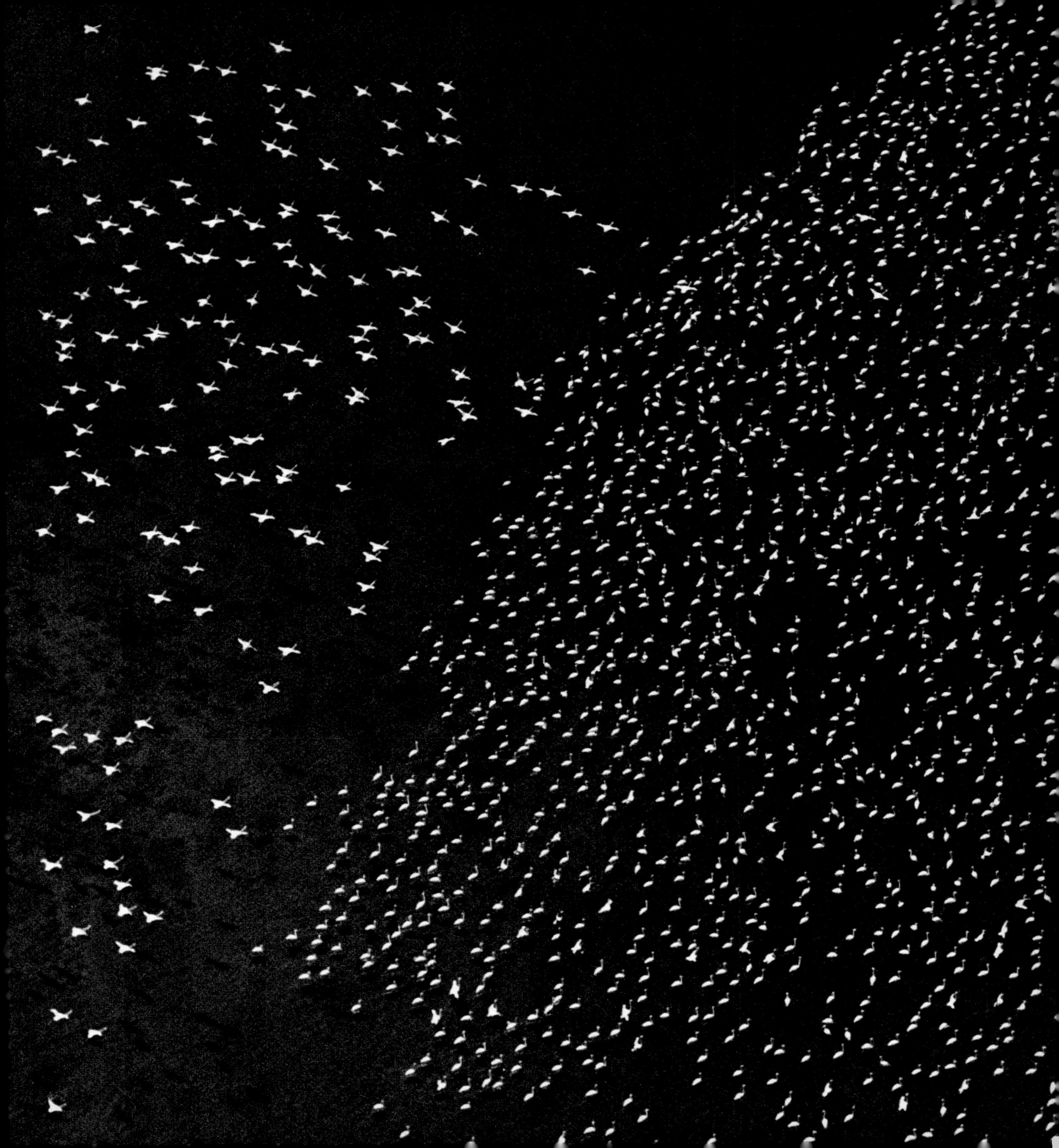

Europe is visited by many migratory birds. They come from the Arctic to enjoy our milder winter, they return from the Equator to our cooler summer. Some fly in parties and flocks and nest alone, others travel almost alone but breed in enormous colonies. For centuries their appearance was the signal for wholesale slaughter and the unlimited plundering of nests. It took a long time for man to turn from bird catcher and egg collector into ornithologist, no longer using cliff colonies of Northern coasts as food stores, but protecting them as natural landmarks. Such colonies are few and far between. Only where there are several favourable factors, such as coastline, climate and ocean currents, and where safety and plenty of food make it easier to rear a new generation, can such massive colonies survive. Today for the first time we understand how all the conditions essential for life are linked together. From now on at least, places which have for centuries been the traditional breeding grounds of a particular species are likely to be left alone, subject only to a moderate cropping of the breeding colonies, which does no harm.

➤ Knots *(Calidris canutus)* gathered in winter. These birds, which breed on the Arctic tundra, winter on European coasts from the Hebrides to Sardinia, where there is a temperate climate.

➤ Gannet *(Sula bassana)* breeding colony on the Bass Rock in the Firth of Forth in Scotland. There are a dozen other colonies of gannets in the triangle Brittany–Iceland–Alesund.

There are two Europes — one for the animals and one for man. Some animals by their very nature set up their own frontiers, spending their whole life in one small area; others roam over the political barriers which divide Europe today to seek the most favourable environment in free flight. For a long time certain birds which like to breed by calm waters have avoided the industrialised West — the ancient banks of streams and inland lakes, in many places undisturbed, of the less developed countries of eastern Europe provide a peaceful haven for them. But despite this division among the animals, familiar landscapes at home or on holiday do not lack animal life. We must learn to look for this, to rediscover nature, to be able to wait, without craving sensation. With patience and understanding we can discover in woods and fields many remarkable forms of life. Whether we see it as a hobby or a science, what makes the exploration of natural phenomena so exciting is the complete independence of the wild animals. Although they have little need of us, we must look to them for the answers to all the thousands of problems of life on earth, answers which we cannot provide for ourselves.

← Cormorants *(Phalacrocorax carbo)* on an island on the Czerwicz lake, Poland, where 400 pairs nest each summer.

→ Over the Danube delta, Roumania: Spoonbills *(Platalea leucorodia)*. *Below:* 1 Glossy Ibis, in silhouette.

As man moves more and more towards a collective society he takes more interest in those creatures which are living in large congregations of like species. One question which occupies students of behaviour among animals is that of individuality where there are hundreds of thousands of sea-birds squeezed into an extremely small space. Therefore field biologists have made a study of the Guillemots and Kittiwakes on cliffs which teem with them in spring. To learn anything worthwhile they must spend many summers among the birds, and absorb the atmosphere of this new world. Man himself learns by trial and error, so he is puzzled by the existence of other animals which depend on a pattern of inherited characteristics. Among the vast crowds of Guillemots gathered on one cliff the members of a pair are able to recognize each other and are concerned *only with each other*, although they build no nest which could become a focal point. The female Guillemot lays her pear-shaped egg directly on the bare cliff. . . close to other eggs of the same species. Hundreds of experiments have been conducted, in which eggs have been moved to different spots and marked hens have carefully fetched their own eggs again and brought them back, clamped between their bellies and their webbed feet. On the 26th day of incubation an interesting experiment can be carried out. If an egg, which is due to hatch in 4 days time and in which the chick is already chirping, is separated from its parents and tape recorded calls of several different parent couples are played over to it, it will only answer the voices of its own parents. . . and conversely adult birds listening to recorded calls only react to the distress call of their own chick. There is such a marked difference of contact in these noisy colonies that parents and young can even find each other again when, several weeks later, thousands and thousands of young birds which are not yet fledged hurl themselves into the raging sea, desperate after one or two days of enforced fasting.

➤ Part of the "Pinnacles" on the Farne Islands off the Northumberland coast, England: Guillemots *(Uria aalge)* and a few Kittiwakes *(Rissa tridactyla)*. The Guillemots prefer high, steep cliffs — North as far as 70° of latitude and South as far as Portugal — so that the fledglings can plunge direct into the sea.

If there is danger from men or gulls, the Oystercatchers *(Haematopus ostralegus)* fly up to utter their strident "pic, pic, pic" of alarm. They breed mainly on the low-lying coasts of Europe, with the exception of the extreme South. Their hunting grounds vary with the tides, but what they dig out of shallows and surf with their red beaks consists mainly of crustaceans and molluscs like cockles and mussels, seldom the delicious oysters which gave them their name and which are very difficult to open. They congregate in very noisy flocks and they fly strongly, beating their wings rapidly.

In autumn, when most of the seabirds have left their nesting colonies, the Grey Seals haul out on their traditional breeding grounds on rocky islands and sandy bays round the northern and western coasts of Britain; but in the separate Baltic population this return takes place in February. Once ashore, these swift and agile swimmers, skilled in the art of fishing, become helpless mountains of fat, and man has shamelessly exploited their weakness. Their survival was seriously jeopardized, because their slow rate of reproduction could not make up for the losses. After years of persecution, the Grey Seal received protection in the British Isles just before the First World War. Today the world population is estimated to be 46,000, some two thousand on the American side of the Atlantic, five thousand in the Baltic, and the rest round the British Isles, Iceland, the Faeroes and Norway. The largest single colony, estimated at 9000 seals, is on little North Rona off the north coast of Scotland. After eleven months' gestation the female gives birth to one single pup, which, although big and well developed, is not nearly as independent as a new born Common Seal, which sheds its furry coat immediately after birth and can swim with the herd only a few hours later. The Grey Seal pups, on the other hand, keep their white curly coats and cling to the land for two perilous weeks. Their mothers' milk is so rich in fat that they put on pounds every day. After being pushed and shoved a great deal, however, they finally overcome their fear of water, for towards the end of the autumn the bull herds together the cows which belong to him and soon afterwards hunger forces their deserted young to learn to fish.

➤ Nursery of Grey Seals *(Halichoerus grypus)* in the Hebrides: a cow is guarding three pups, one of them trying the water in the small pool for the first time. On shore five other adults are basking and three more are playing in the surf.

⅄ Common Seal *(Phoca vitulina)*, the most numerous species in the Northern Baltic seas and Atlantic coastal waters from Murmansk to France and Portugal.—Only the Common Seal and the Grey Seal regularly inhabit British waters. All seals are fish-eating, have flippers which act like a fish's tail, to steer, and have huge eyes, which can see under water in the dark, as well as sensitive whiskers.

A school of Pilot or Caa'ing Whales or Blackfish *(Globicephala melaena)* in the North Atlantic.

At least 50 million years ago, at the same period as the ancestors of the seals, the forerunners of the whales turned back to the sea, to become the only group of lung-breathing, warm-blooded animals to spend their whole life among the fishes. There are the Baleen Whales which feed on plankton, and include the largest of all such as the Blue Whale (80 to 100 ft), which weighs up to 150 tons; there are record divers, like the Sperm Whale, which fight the Giant Squids at a depth of some 1,500 ft; greedy predators, like the Killer Whale, a single one of which had devoured 14 Porpoises and 12 Seals before it was caught. These killers are the terror of the sociable Pilot Whales *(photo)*, which panic and take flight, often to be led astray by their leader and run aground on the coast in large numbers. This herd instinct is made use of by the whale-catchers of the Faeroe, Orkney and Shetland Islands, and even today they still set out in open boats to drive herds of Pilot Whales into their bays. But 200 years of heavy, gradually mechanized whaling have meant that the old European hunting grounds are exhausted and the Greenland Right and Biscayan Right Whales have been almost exterminated. And so, as well as making a large annual catch in the Faeroes, the whaling fleets with their factory-ships also travel to the Antarctic seas where the number of whales caught annually is now limited by international agreements.

The White-beaked Dolphin *(Lagenorhynchus albirostris)*, which accompanies steamers in the North Sea.

The dolphins, which are very small in comparison with the other Toothed Whales, 8 to 9 ft. long, have sustained the smallest loss. They have all the advantages of the whale's constitution plus a network of veins which store oxygen, allowing them to remain under water for up to 15 minutes. Their cornea serves as a water-pressure gauge, preventing them from diving too deeply. The trachea opens directly into the glottis so that they are incapable of swallowing the wrong way, and their larynx can produce unusual whistling and chuckling noises: they use these to communicate and guide each other to the surface when there is danger. Along with many other Toothed Whales they also emit ultrasonic waves to locate surrounding objects by means of echoes. The dolphins and the young of all whales are born under water, tail first. They are already fully developed (the baby Blue Whale measuring over 23 ft.) and swim at once to the surface to take in air. The mother swims on her side so that the young can be suckled while she is swimming and can breathe at the same time. This sight is a public attraction in the dolphin seaquariums in Florida; these highly intelligent animals perform many remarkable tricks, and some zoologists believe they could be taught the rudiments of human speech and be used as fish "shepherds".

Off "King Karl Land", Spitzbergen: Polar Bear *(Thalarctos maritimus)* with her eighteen-month old young.

This peaceful picture of the Far North is a deceptive one. It is the sort of scene viewed by parties which go on motor boat trips organised by Norwegian travel agencies, to hunt Polar Bears in their search for trophies. But these trips which take place in ice-free coastal waters mean death only for females and their young, which are suckled well into their second year of life and rely for food on the birds' eggs and berries of islands free from snow, and on the leaping salmon in the river mouths. Polar Bears also eat Arctic Willow, mosses and some crustaceans. In summer, however, the male Polar Bears travel far out into the belt of drift-ice, to find the breeding grounds of the seals. They have a thick layer of fat and can stay in the water for several hours. Only the Eskimos hunt these male bears, as they have done for centuries, sparing the pregnant females to be killed by "civilised" man purely for entertainment. This has meant that in the last 40 years the Polar Bear population has been halved. Of no avail are the precautions taken by the female, who, defying the dangers of the Arctic climate, digs a den in the snow with an entrance tunnel up to 3 yards long where she bears in late winter her two blind young, which weigh only 25 oz. each. The small family leaves this glacial hiding place in March.

On the shores of Shillay, Hebrides: encounter between rival Grey Seal bulls *(Halichoerus grypus)*, up to 9 ft long.

At one time Polar Bears drifted with the ice as far south as Iceland, solitary Walruses ventured as far as the North Sea, and Grey Seals were found right down to Rugen. Today the Grey Seals' most southerly quarters on land are on the coast of Cornwall. In autumn the oldest bulls swim on ahead of the females and the one-year-old offspring, inspecting the coast suspiciously. When the pregnant females arrive, fierce fighting breaks out between the males as they strive to collect a harem, although the cows are reluctant and the bulls cannot approach them until this season's pups have learnt to swim.—The different sub-species of Seal have very different life cycles: the Harp or Greenland Seal, from the central Arctic, make for the Jan-Mayen Sea between Spitzbergen and Iceland in their thousands in April, but those living in the eastern Arctic congregate in February in the White Sea below the Kola peninsula. By these icy waters seal-catchers lay in wait: Nansen tells us that in the nineteenth century they boasted of killing and skinning 300 young Seals each daily. Today, 32 islands in the White Sea are protected by law as a part of the large Soviet Kandalaksha Seal-breeding reserve. Migrating Harp Seals swim in single file, resembling a giant Sea Snake.

European Elk *(Alces alces)* bathing. A reminder of the herd of Elks in the Courland district of Memel which was wiped out in the Second World War. See also pages 212, 213.

The ancient species of the Elk, which is today the largest deer in the world, has never been outdone in courage except by the giant Irish Deer of the Ice Age, whose projecting palmed antlers, 8 ft long, have survived for thousands of years. The Elk once covered an area comprising the whole of Northern and Central Europe as far as the Alps, but has now withdrawn into the marshland forests of Poland, Scandinavia and Russia, where a single 12 point bull rules over a thousand acres. Wary and independent, the Elk stag sets out for pasture in the early morning, spends the middle of the day ruminating drowsily, and only towards evening does it begin to make its way back again with its springy step on cloven hoofs through the marshland, feeding on young reeds and breaking off the juiciest shoots of the deciduous trees. It is known to gnaw at the bark of trees and also eats grass, leaves and water plants. In September it extends the limits of its territory. The rutting impulse drives it into unknown regions, swimming across streams and inlets, and crossing cultivated land by night. If it meets another stag with the same purpose, the two enraged giants rear up and cudgel each other with their steely forehooves. They fight to gain a single cow, for they are the only deer which do not gather together a harem, preferring to wander off again after a few days. The cow will bear her two calves in early summer.

A herd of reindeer *(Rangifer tarandus)* on the last stage of their great spring migration: they aim to swim to the greenest island off the coast of Finland.

The Ice Ages drove the wild reindeer southwards, as far, even, as the Mediterranean coast. In the period between the Ice Ages, however, these, the most sociable of all deer, returned towards the North as it became green again. The 3,000 years old cave drawings at Bohuslän in Sweden depict reindeer hunts, but the huntsmen became herdsmen, seeing in the enormous herds not only supplies for today and tomorrow, but property for their children and their children's children. However, the reindeer forced its own rhythm upon the herdsmen, and as a result, today 30,000 Laplanders are still true nomads. The routes they take are determined by the leading deer of the herd, and every spring and autumn men and animals travel hundreds of miles over tundra and taiga from the forest region to the green coasts and back again. The calves are born on the way there, and on the way back the stags fight for the reindeer does, who also bear antlers. Now the herdsmen must be doubly watchful, for often defeated stags lead whole groups of hinds astray over the boundary of one country or another, in order to found their own herds. In Southern Norway, owing to measures of conservation, the only surviving wild European Reindeer have greatly multiplied. Reindeer occurred in Scotland in Viking times. Recently there has been an attempt to re-introduce them.

Musk Oxen *(Ovibos moschatus)* in the Far North, in defence position. Four strong males stand at the corners, and the cows and calves huddle together in the middle.

When fossilised remains of Musk Oxen of the Ice Age were found in Central Europe it was believed that they had died out with other animals of that period. When the glaciers dispersed, their shaggy coats became drenched with water during the prolonged rains that followed, so that many thousands, who had survived the dry icy winds, died of pneumonia. However, in the 18th century, trappers from Northern Greenland brought the news that they had come across shaggy oxen, which formed the characteristic defence position of the Musk Ox at the approach of a human being. The Musk Ox is now the world's most northerly ruminant. It spends the short polar summer in small herds in fields and meadows, living mostly on grass and moss. But in the winter herds of about 60 animals crowd together, wherever they can find vegetation free from snow. Sometimes they spend whole days without food. The characteristic formation which protected the young from storms, wolves and polar bears, was useless against shooting by human beings. Thus thousands were slaughtered and they are now only to be found in the north of Canada and North Eastern Greenland, where they live as protected animals. The Musk Ox is over 8 ft. long, with a very short tail. They have many anatomical resemblances to sheep, including the broad horns. As well as grass and moss they feed also on arctic willow, grazing during both day and night and only stopping now and then to chew the cud.

Herd of Bison *(Bos bonasus)* in the spacious reserve of Bialowicz, near the marshy source of the Narew, where wild horses, the Tarpans, used to graze until 1880.

In the famous cave paintings of Altamira and Lascaux wild cattle and Buffalo feature largely, often being depicted as having been caught. It was not until thousands of years later that man used dogs to help track and herd Bison and Aurochs. These ancestors of European domestic cattle were first tamed 8000 years ago; they were driven out of their pasture in the 17th century, and the herds of Bison dwindled until the Russian Czars ensured them a last refuge in the forest of Bialowicz. But, out of 700, few survived the 1st World War and the chaos of the October Revolution. Today the situation has improved somewhat and more Bison make their way through the forest and heath of the Polish reserve. In summer the huge steers leave the herd, but they return in the autumn to find a mare. The steers have symbolic fights with bushes and tree trunks and then rush towards each other, bellowing. Forehead to forehead they froth at the mouth and stamp, repeating this again and again until one of them retires. The opponent is not killed but simply leaves at the first sign of weakness. Bison feed on grass and branches of trees. They are up to 11 ft long and the bull is a great deal larger than the cow. They have shaggy hair and a fatty hump; their thick winter coat comes off in large tufts in the spring. The European Bison or Wisent is closely related to the American Bison, which is commonly known in North America as the Buffalo.

Of all the European terrestrial predators the Badgers, which are very cautious animals, spending almost the whole daytime in their setts, have best survived the invention of gunpowder. It has even been to their advantage, since their natural enemy, the Lynx, has been largely exterminated by shooting in the whole of Western and Central Europe and now survives only in Northern Scandinavia, Eastern Europe and the West of Asia. The Lynx was pursued ruthlessly because it preyed on such animals as sheep and deer, often killing more than it was able to devour in a night. Earlier the Lynx acted as a scavenger, simply killing old or dying animals, leaving what it did not eat for smaller carnivores. During the day the Lynx hides in dense scrub and in crevices in rock, hunting in the evening. It hunts birds as well as mammals, and some rodents. Hares are one of its main targets. The female gives birth to 2 or 3 young in May or June and stays with them until she has the next litter, when they are nearly mature. Lynx vary in colour, from yellow through red and brown to grey and very pale grey in winter. They also have spots which may be placed in many different positions and may be many or few. They have straight ears and ear-tufts 2 ins long, and their tails are short and tipped with black. There is also a somewhat smaller species, the Spanish Lynx, which has more spots on its coat and is only found in Spain and Portugal.

➢ Lynx *(Lynx lynx)* preparing to spring. A strong male weighs up to 100 lbs and is well over 1 yard long.

⮘ Badgers *(Meles meles)* returning in the early morning from a nightly hunting expedition, searching for insects, frogs and small rodents. In autumn they also eat berries and crops of wheat. Their setts often extend for more than 30 yards and consist of a series of passages, chutes and burrows, going down to a depth of over 15 ft. Setts are often passed on from one generation to another and sometimes house a fox family too.

In the 17th century there were still Beavers *(Castor fiber)* on all the rivers of Europe — and today place names such as Biberist, Bevers, Biberach, Bebra, Bièvres still remind us of these animals. They were hunted for several reasons: because of the damage they did to the trunks of trees, because their pelts were prized by the nobility, and because the "castoreum" was used by apothecaries. They were also eaten during Lent. Beavers are now found along the Rhône upstream as far as Avignon; at the confluence of the Mulde and the Elbe; in Southern Norway, Finland and along the Don. In Sweden, Austria and Switzerland attempts are being made to resettle them in pairs. Beavers build their dwellings with mud and sticks on river banks, and then construct a dam downstream which causes the water level to rise and make a pool around the dwelling or lodge. Although Beavers feed mainly on bark, they also find the juicy shoots and leaves of newly felled trees appetising in summer, and store branches on the bed of the river or lake to supply them with food in winter. In early May the female bears 1 to 3 young the size of rats. The adults are the largest rodents of Europe, being 40 to 52 ins long, including the tail, which is flat and covered in scales. The toes of their back legs are webbed. There have been no Beavers in Britain for perhaps a thousand years, but a form related to the European Beaver inhabits North America. The activity of Beavers has always been beneficial in uncivilised areas where there are forest fires or drought and where they can re-create fertile land from barren, thus providing sustenance for many other animals of their habitat.

At the beginning of this century the fur producing animals of Europe had almost been exterminated. Therefore in 1905 a Bohemian prince introduced a few pairs of the North American Musk Rat *(Ondatra zibethica)* in Dobrisch, near Prague. Thus he was responsible for the Musk Rats quickly becoming a pest all over Central Europe, building their burrows in the banks of rivers and canals. After their establishment in Britain a determined campaign exterminated them by 1939. In Europe, within two decades, they had taken over 125 thousand square miles of cultivated land. In France, Belgium, Finland and Switzerland they also became a pest and all possible steps were taken to prevent the building up of new colonies. This is difficult, because one Musk Rat pair has 21 offspring a year and these are themselves pairing a few months later. Like the Beaver, the Musk Rat has underwater entrances to its burrow, but unlike the Beaver it does not cut wood, but uses reeds for building, filling in the cracks with mud. The Musk Rat also has webbed back feet and a scaly tail and is found at the edges of lakes, rivers and fishponds. Water plants are almost its entire source of food. It is 22 ins long, with a thick reddish-brown coat, and a waterproof undercoat. It is a nocturnal animal and does not travel far, except when migrating in September, when it will attack cattle or man if threatened. It is a rapid swimmer and can travel as much as 100 yards under water before surfacing. The name Musk Rat arises from the scent secreted by scent glands near the anus, at the same time as the dung.

Very few Old World Mink *(Mustela lutreola)* can still fish freely on rivers, streams or swamps of North Eastern and South Eastern Europe today... for with their fine, glossy fur they have little chance of escaping capture. The Sable has also vanished from Northern and Eastern Europe, and its breeding is now a Soviet state monopoly, the soft silky pelt being valued according to quality at 2,000 to 8,000 roubles. However, in the absence of European parent stock, the Mink farms of Britain and Western Europe breed American Mink: for one high-fashion mink coat 210 animals are killed, and the layer of fat under the skin which gives the coat of the live Mink its silver sheen, is manufactured into beauty cream. Recently variations in colour such as pearl and platinum have been produced by mutation. Mink in their wild state feed on frogs and crustaceans, and even water birds or rodents, as well as fish. The female bears 5 to 6 young at the end of April. Mink are from 19½ to 21½ ins long with swimming webs between their toes. If anything frightens them their anal glands produce a repulsive stench. Occasionally the American Mink, which are bred in Mink farms, escape, and there have been reports of some of these living wild in Devonshire. These American Mink are larger than the Old World Mink and are closely related to the Polecat and Weasel. They swim well and are mainly found among reeds or the undergrowth near water.

Although Otters *(Lutra lutra)* are found all over Europe and into Asia, they can be seen in any numbers only during the daytime in the rivers and lakes of Scandinavia. For up to 70 years ago in the other countries of Europe they still fetched a high price when shot as pests. There are recipes in old cookery books for using the black flesh of the Otter in the preparation of a delicious dish for Lent. Now, in those countries where it has become rare, the Otter is protected by law and can fish unmolested by human beings or Otter hounds. As well as fish, it will eat molluscs, frogs, small rodents and even water birds, and has special places on the land where it takes larger fish to eat them. The Otter enters the water at about 2 months old. The web membranes between its toes help to make it a very capable swimmer, its long tail being used for steering. It has the means of closing its ears and nostrils, thus facilitating diving, and its fur is so watertight that its thick undercoat does not even get damp. The Otter can manage very well on land and will wander over great distances, especially during hard winters or in order to find the rivers where the elvers run. It usually mates in February, with much splashing and playing. 2 to 3 young with grey woolly fur are born in the grass or moss-lined nest in the holt, usually in early spring, though they can be born in any month. They are suckled for 6 months.

Those Salmon which manage to struggle up-river through rapids and waterfalls, and circumvent weirs and dams on their long detour up the fish passes, have left their greatest enemies behind them: they have escaped the seals (and, in parts of Europe, the Dolphins) which lay in wait for them on the estuary of the river, and have also avoided the fishermen's nets stretched out across the river. True, the fishermen know the breeding season of the Atlantic Salmon which ascend British rivers very early in spring, and appear in the Rhine at the end of July to start their 600 mile journey up-river... yet still ichthyologists do not know where these nomadic fish go, when, having been born in the mountain streams of Europe, they disappear into the Atlantic Ocean as sardine-sized "smolts". All that is known is that, as sexually mature fish 2 to 4 ft in length and 3 to 6 years old, they return up the stream to where they hatched. It seems unlikely that they remain inshore, near the mouth of the river they have just left, for Salmon are very seldom caught in the dragnets of fishermen in coastal waters, though they are sometimes found in drift nets. The most recent theory is that the two year-old smolts from the Northern European rivers make in shoals for the water under the pack-ice of the Arctic, where the deep waters, rich in mineral salts, contain an abundance of tiny organisms, crustaceans and fish. However, after 2 to 4 years the now full grown Salmon are driven to set off again by some mechanism which is still not thoroughly understood. Nevertheless, mass marking of the young fish has clearly demonstrated that, thanks to their excellent senses of smell and taste, most of the surviving Salmon find the mouth of their own river again and even the place at which the stream where they were born branches off. After travelling this far the Salmon go off in pairs. The female lays 10,000 to 20,000 eggs the size of peas in beds of gravel. These are then fertilized and carefully covered. From the time they enter the river, the fish eat nothing at all and when they finally head back towards the sea they are thoroughly exhausted and emaciated "Kelts". Very few of them will ever return alive and only rarely does a Salmon survive to spawn three times. Meanwhile, in the gravel beds of the mountain streams the young hatch after 3 to 5 months, and live off their vitelline sac for 6 weeks. They then hunt for food in fresh water for one to six years, until they also are driven towards the sea, which only 10 per cent will ever reach. Later only one per cent return to the breeding grounds.

The Salmon *(Salmo salar)*, up to 5 ft long and 50 to 70 lbs in weight. It has to travel further than all the other fish which have their hunting grounds in the sea and their breeding grounds in fresh water. Shad, Sturgeon and Sea Trout are satisfied with spawning in the lower reaches of the river. Only the River Lamprey penetrates as far as the Salmon, accompanying it as a blood-sucking parasite right to the mountain source of the river.

From late summer until early spring the Puffin has its habitat mainly on the Atlantic ocean, sometimes in coastal waters, though rough weather may drive it inland or towards the calmer waters of the Gulf Stream. Birds which breed on the coasts of Britain have been sighted off Newfoundland; other groups of birds make their way southwards towards the Straits of Gibraltar. But, wherever they go in the winter, in March crowds of Puffins return to their traditional breeding grounds on cliffs and islands off the coasts of Iceland, the Faeroes and Norway, the British Isles and Brittany, where thousands of them lay their eggs. Puffins may dig a tunnel, using bill and feet, in which the female lays her single egg, but sometimes she uses an empty rabbit or Shearwater burrow. The chick does not hatch for about 6 weeks. One parent bird always stands guard while the other goes fishing. The very unusual bill is especially useful for this purpose: as soon as a fish is caught it is pushed to the back of the beak, leaving the front free again for more fish; only when the bill is full does the Puffin fly back to the cliffs. At 6 weeks none of the young can fly, but from the moment they enter the sea they can swim and dive. Not for several more weeks do they actually fly up into the air and they will not reproduce until they are at least three years old. The bills of the young Puffins are brownish or blackish and in the adult Puffins both bill and legs change colour in summer and winter. The bill has more yellow in the winter and the legs, which are red in summer, are entirely yellow in winter. Puffins perch in an upright position, but rest horizontally. During the breeding season their voice produces a low growling "arr".

Puffins *(Fratercula arctica)*, 12 ins long. Their greatest enemies on land are Skuas and man. On "Draga Lund" in the Faeroes tens of thousands of brooding birds are pulled out of their nesting holes every year by sticks with hooks on the end. The breeding colonies of the Puffin are larger than those of the Guillemot, which it resembles in behaviour, except for its nesting habits.

One of the biggest northern sea-birds is the 36 ins long Gannet *(Sula bassana)*. See also pages 5 and 73.

With a wing-span of over 6 ft the Gannet glides majestically over the seas of the north, and in the winter it sometimes roams as far southwards as the west coast of Africa. With nostrils closed and ear openings narrowed it dives after fish at a great speed from heights of 100 ft or more and to a depth of up to 6 ft, and often devours its prey while still under water. For centuries fishermen and people living near sea-shores, and even today on the Faeroes, have been killing Gannets on their densely populated breeding grounds, which are to be found on steep cliffs around the British Isles, north to Iceland and east as far as Norway. Now, however, the killing is done with restraint, as the Islanders have come to realize that these colonies of Gannets can later provide food for their children and grandchildren. The young Gannets are brown in colour, gradually developing the white plumage with long black tips on the wings of the adults, which also have a yellowish head and neck. In order to rise into the air from the water Gannets need to take off flapping heavily and they are awkward on land. However they swim very well. The family of birds to which Gannets belong is called the Booby family, and is related to the Pelicans. They are often seen far out at sea and seldom venture inland unless they are driven there by storms.

The Long-tailed Skua *(Stercorarius longicaudus)* measures 21 ins up to the tip of its tail. Wingspan 40 ins.

Skuas are renowned for their piratical behaviour. They themselves do not have the thrust to dive after fish, so they circle over the water and watch to see what the Puffins or Gulls are hunting. When a hunter catches a fish the Skuas pursue, harrying it continually with their skilful flying until it finally disgorges its prey. They then seize the prey in flight, devour it hastily and immediately start looking for more sources of food. The Long-tailed Skua lives mainly offshore, but also on pelagic waters, and breeds in small scattered colonies on the high fells and marshy tundra of Northern Scandinavia. There the female lays one or two eggs, and when the young hatch the gangster activities of the parent become more frequent and widespread. They will even prey on the young of other species, except in the very near neighbourhood of their own breeding grounds. All Skuas have tail projections, but the Long-tailed Skua, which is actually the smallest of the Skuas as far as bulk is concerned, is much longer than any of the others, because of the extraordinary length of the central feathers, which extend from 6 to 10 ins beyond the rest of the tail. It flies very gracefully and swims with neck stretched upwards and tail cocked. It is usually silent except during breeding, when it cries "kree".

The Osprey is the only member of the *Falconidae* which preys exclusively on fish. It hovers and circles over fresh or salt water, examining the surface yard by yard. As soon as it sees a fish it hurtles towards the water feet first, with its wings folded against its body. It brakes just above the waves, stretches out its claws, plunges in and seizes the fish. Its talons are supple, with prickly, horny scales, enabling it to grasp its slippery prey. It then flies back to the nest, which is an enormous construction of driftwood and sticks torn in flight from dead trees, on a cliff top or small island or in trees or ruined buildings. When the male brings a fish to the nest or eyrie, the female removes the entrails and fins and gill cover and feeds the young with the flesh. The Osprey is brown above with white underparts, the head being white, with a brown stripe through the eye. It produces a sort of shrill whistle, like a young bird. It is occasionally to be found breeding in scattered groups. It recently returned to Scotland and nested successfully in Inverness-shire 1959–62.

The Osprey *(Pandion haliaeëtus)*, length 20 to 23 ins, wingspan up to 5 ft, is mainly a summer visitor to its breeding grounds in Scandinavia and Eastern Europe. It winters in the Mediterranean, where it is also a permanent resident.

With a wingspan of over 6 ft and the primaries spread out wide, eagles circle higher and higher and soar majestically above their terrain, which they may occupy for decades. The territory of the White-tailed Eagle or Sea Eagle encompasses great stretches of water and the heathland around its huge eyrie in some tree or on top of rocks. Sometimes small birds nest safely in the foundations, for the White-tailed Eagle seeks larger prey: fish, ducks, water fowl and even weak mammals as large as Roe Deer. The Golden Eagle used to hunt over ground rich in wild life, but centuries of senseless pursuit and killing have driven it into the most isolated mountain valleys, although it occasionally inhabits cliffs or plains. It soars for miles over valleys and Alpine peaks up to heights of 10,000 ft, in search of its favourite prey, the Marmot. But during the winter it hunts Ptarmigan and Mountain Hares on the sides of mountains and in snow-filled gorges. It also feeds on carrion, such as the corpses of sheep or chamois which have died from disease or accident. It calls very infrequently, producing an occasional harsh yelp and whistling sounds.

← The White-tailed Eagle *(Haliaeëtus albicilla)* still breeds on the coasts and lakes of Northern Germany, Scandinavia and Iceland, from the Baltic to South Eastern Europe, and on Corsica and Sardinia.

→ The Golden Eagle *(Aquila chrysaëtos)*, 30 to 35 ins long, builds its cliff eyrie on the hills of Scotland and above the mountain forests of Central Europe and the Balkans.

Short-toed Eagles *(Circaetus gallicus)* in the eyrie. Their breeding grounds today are in the triangle Spain–France–Italy.

The Short-toed Eagle is 25 to 27 ins long, with a large head not unlike that of an owl. It has little apparent interest in feathered or furry prey, but is skilled in catching snakes and other small reptiles and amphibians. When one of these birds, soaring over ravines and mountain slopes, discovers a snake, whether a harmless Grass Snake or a poisonous Viper, it seizes it with its sharp beak and carries it to the nearby eyrie in a tree, where a single young bird will have hatched after five weeks' brooding. The parent bird grinds the food ready for the fledgling to swallow, but soon the tiny bird will devour whole lizards and blindworms on its own. However, it will be a full three months before it dares to leave the nest for the first time, when it will prey on snakes in its turn. The young bird has brownish underparts which will change to the almost even white of the adult bird, which is grey-brown above with blackish wing-tips or *primaries*.

Spanish Imperial Eagle *(Aquila heliaca adalberti)* with young. Length 33 to 35 ins; wing-span 79 ins.

The Imperial Eagle lives in the extreme South West and South East of Europe. In Spain and Portugal it preys on rabbits and birds, and even feeds on carrion; in the Balkans and South Russia it hunts above all the Souslik, the Ground Squirrel of the steppes. When first hatched, the nestling weighs only 3 oz; a week later its weight has been quadrupled and in 30 days it weighs 6½ lbs — nearly as much as the adults. As in all diurnal birds of prey, the chick's eyes are open from the first, but the lids are kept shut for two weeks. The young bird has yellowish brown down; in the adult the plumage is dark brown with a pale crown and throat. The Spanish form has white shoulders which distinguish it from the typical form *(Aquila h. heliaca)* inhabiting South East Europe. As darker feathers appear in its plumage, the fledgling becomes more and more restless, shaking its wings and springing up and down on the nest, until at last it can fly. The eyrie is built in a tall tree in open country.

Griffon Vulture *(Gyps fulvus)* with the corpse of a fawn, Cota Doñana, South Western Spain.

Where there are vultures to pounce upon carrion, dead and decaying animals will never become a breeding ground for vermin or infection. In vast areas of the tropics vultures still provide the only substitute for sewerage and sanitation. However, in Europe today there are only a few regions, where the Griffon Vultures circle over the villages when slaughtering is taking place: they still breed on rocky ledges in Spain and Southern France, Sardinia and Sicily, Greece and Roumania. During the breeding season they croak and whistle at their nest sites. They nest and roost communally and also fly in parties, haunting all types of country, especially the mountain ranges. Soaring at great height for hours without moving their wings, the vultures appear wherever some wild animal has died, a sheep has fallen off the rocks or a cow has succumbed to some disease.—All vultures have longer wings than eagles, and shorter tails. The Griffon Vulture is about 38 to 41 ins long and has a very short tail. This and the wings are dark in colour, the rest of the plumage being sandy except for the head and neck, which are white and downy.

Enough of the Griffon Vulture's prey remains to feed the Raven *(Corvus corax)* and Black Kite *(Milvus migrans)*.

As soon as one Griffon Vulture alights near an animal's corpse, it is joined by others of its kind, while Raven, Kite and Egyptian Vulture may also assemble to share the feast. In half an hour the carcass will be completely stripped of flesh. Then the Bearded Vulture may arrive, using its steel-hard beak to crack ribs and the larger bones; it devours even hooves, jawbones and teeth and eventually flies off with its huge crop filled to bursting... very strong digestive juices break down the bones and extract marrow and jelly from them. Although the Egyptian Vulture is not generally sociable it sometimes joins other scavengers for the final pickings. The Black Kite is much more sociable and often appears in flocks; besides feeding on animal remains it will devour dead fish. This species often nests communally in trees. The Bearded Vulture is found mainly in Greece and the Egyptian Vulture in Spain, Italy and Greece. The Black Kite, on the other hand, ranges all over Southern and Central Europe in summer. The Raven is resident throughout Europe except for Italy and Central Europe north of the Alps. It also inhabits Iceland.

The Bearded Vulture's German name *Lämmergeier* shows how much it was held in detestation by shepherds. Not 100 years ago, it was virtually exterminated in the Alpine regions. Ornithologists have tried to re-establish breeding pairs there recently and a few are now to be seen in summer in the Eastern Alps. A very few *Gypaetus barbatus* still nest in the mountainous parts of the Iberian Peninsula and the Balkans; yet they abound in the mountains of Africa and Asia. These majestic vultures have a wing-span of 10 ft. They are usually solitary and more active than other vultures and for survival they must be able to find enough carrion, for when they have to prey on living creatures, for every lamb or kid they kill they will be hunted for their lives.

The little Egyptian Vulture *(Neophron percnopterus)* is a distinctive feature of the landscape in Southern Europe. Where larger scavengers gather to devour carrion it will wait patiently for a share of the offal. It nests on cliffs and trees and inhabits all types of country, including villages. It is only about 2 ft. long and has a very slender bill compared with other vultures. ➤

The Bee-eater is using a skull as a look-out in the bare countryside, where it watches for bees, wasps and hornets, ready to dash off in pursuit at tremendous speed. It is the most vividly coloured of European migratory birds and yet it does not hesitate to soil its brilliant plumage in heavy tunnelling work. Like the Sand Martins, the highly gregarious Bee-eaters bore their nesting holes in sandy slopes and river banks where their eggs and young are protected from marauders. It takes them 2–3 weeks to dig the 5 ft long tunnel and as in the case of all birds, their strong curved beak is their only tool.

The beaks of birds have to do all the work of the "lost forepaws": according to the species, they serve as chisel, hammer, scissors, spoon, sieve, sucker, pincers or nut-crackers. They are used for feeding, building the nest and for defence and this versatile tool is the result of millions of years of evolution. This began with a small reptile with membranes extending from the shoulder, enabling it to move in short gliding jumps. Eventually this membrane extended as far as the claws and the forefeet were superseded by wings. The neck became very supple and the jawbone gradually became modified until the beak was produced. Then again millions of years were to pass before the first feathers sprouted from the scaly skin and before gliding developed into the fluttering of a new, feathered class of animals.

➤ The Bea-eater *(Merops apiaster)* visits Southern Europe regularly in summer and has wandered as far North as England; once it even tried to breed in Scotland. The sexes are alike in colouring and the birds are 11 ins long. They glide gracefully, rather like Swallows, and during flight they constantly produce liquid sounds, sometimes a little throaty. They favour open country with bushes, some trees and poles.

➤ *Pterodactylus scolopaciceps*, the small reptile with the beginnings of wings and the beak armed with little teeth, half scale. It was one of the original ancestors of the birds and glided between giant ferns and palm trees 165 million years ago, when central Europe was still an island between the North Sea and the ancient Mediterranean.

Kingfisher *(Alcedo atthis)*. The bird is leaving the water after catching and swallowing a tiny insect larva...

There are 87 species of Kingfisher in the world, of which the smallest is now found as a partial migrant all over Europe except for Norway, northern Sweden and Finland. Many do not leave the streams of Central Europe for the coasts except when they are covered with ice. In April the birds return to their old homes and drive out strangers by whistling shrilly. If their breeding tunnel has been blocked they dig another at least a yard long. From time to time during the digging one of the pair shoots out of the bank and dives into the stream. It will then emerge and perch on a stick or stone where it grooms its feathers, which are oiled frequently with the secretion of the coccygeal gland, to preserve the brilliant azure blue or blue-green colouring of the waterproof plumage and the chestnut underparts. The tunnel in the bank slopes upwards and is widened at the far end to make a spacious nesting chamber, lined only with fishbones which the birds disgorge. As well as fish the Kingfisher also feeds on insect larvae which it finds in the water. It flies very fast, whirring as it does so, and sometimes hovers above the water before plunging after prey.

. . . but as soon as it catches a fish it will fly direct to the nesting hole in the nearby bank of the stream or river.

Both male and female Kingfisher brood for 3 weeks, after which the 6 to 7 blind naked young hatch and are fed with tiny molluscs. However, they are soon able to swallow the small fish which the parents bring head first ready to be devoured. Without stopping at the entrance, the adult birds shoot straight into the nesting tunnel one at a time. After 25 days the tunnel becomes too narrow for the young birds and one after the other they emerge covered with bristles and as prickly as hedgehogs. As they shake the earth out of their stubby feathers, the sheaths which are peculiar to Kingfishers and Bee-eaters are shed, to reveal their magnificently coloured plumage. Soon they in their turn will perch on some vantage point along the stream, where they can look out for prey and from which they can plunge straight in after it. As well as the streams where they breed, Kingfishers also frequent canals and lakes and sometimes estuaries. They often breed in sandpits a mile or more away from the water. During courtship the birds twitter and their call is generally a shrill piping "chee". Their song consists of a short whistling trill, similar in quality to the callnote.

A great deal of technical skill goes into the construction of the intricate nests of the songbirds. Year after year new nests must be plaited and woven, lined and plastered and a suitable site must be found. Every species knows instinctively how to build its characteristic type of nest when the breeding season starts. In April male and female Fantailed Warblers start fetching plant fibres, fluff and spiders' webs and begin building their deep nest, the shape of a Hock bottle, between tall rushes. The stalks to which it is fixed are woven into it and as these grow they will lift the nest up higher and higher. It is protected quite efficiently, usually on the edge of marshland, and in one summer there will be three broods of 4 to 5 young. This high rate of reproduction helps to offset losses sustained by the many Fantailed Warblers which build their nests in cornfields or meadows where mowing destroys them. These Warblers live in both marshland and dry fields or plains. They are resident in Southern Spain, Italy and Southern Greece and the Mediterranean Islands between Italy and Spain. They are the smallest of all the European Warblers.

Fantailed Warbler *(Cisticola juncidis)* at the nest. This Warbler is only 4 ins long, with a short stubby tail, tipped with black and white. It is generally secretive.

The female of the Great Reed Warbler is responsible for building the fine nest on stilts — the male sings meanwhile, repeating each note 2 or 3 times, e. g. "karra-karra", at the top of its voice. The female moistens the blades of the reeds in order to make them pliable, before plaiting them and sticking them together with saliva. Weeks later the nest is finished and has been fastened securely to the reeds on which it is suspended over the surface of the water. In June the female lays 4 to 5 eggs. Only one brood is reared each year. The colonies of the Great Reed Warbler are fairly safe among the reeds ...but sometimes cuckoos lay eggs in some of the nests. The young of these Warblers hatch after 14 days. Soon, when their plumage has become thick enough, they leave the nest and leap about among the reeds, and by the sixteenth day of life they begin their attempts at flying. They swoop low with tails spread out, and often perch in trees or on telegraph wires. They spend the summer in all parts of Europe with the exception of Great Britain and the Scandinavian countries. However they have occasionally been known to wander to England, Ireland and Norway.

Male Great Reed Warbler (*Acrocephalus arundinaceus*) feeding the brooding female. 7½ ins in length, they are the largest of the Reed Warblers.

Many birds nest in holes and empty Woodpeckers' nests provide satisfactory breeding sites for a large number of such birds. However, when the Nuthatch moves in, it actually converts the old nest into a suitable domicile for its own young to hatch, without any danger from enemies. One bird defends the territory, while the other searches for lumps of mud. Using its saliva it plasters the sides of the entrance and any cracks with this mud. It has to make many journeys to complete the work. The entrance is then so narrow that only the Nuthatch can squeeze itself through and the dormouse is kept out. The female then lines the unusual nest with the scale of bark. The eggs are incubated for 16 days and soon there are 7 to 9 hungry fledglings to feed. In order to get food for them the adult birds climb the trees both upwards and *downwards*, unlike any other bird. After 4 weeks the young Nuthatch emerges from the nest and is straightway capable of acrobatic feats. Its large strong feet are very helpful in this. It lives mainly in deciduous trees and is sometimes to be found in winter among foraging parties of Goldcrests or Tits. It is blue-grey above and has buff or chestnut underparts.

Nuthatch *(Sitta europaea)* feeding young. The adult is 6 ins long, ¾ oz in weight. Nuthatches are resident throughout Europe, with the exception of the far North.

Although many Tits are content to nest in any available hole, such as in nesting boxes, watering cans or old tins, the tiny Long-tailed Tit takes from 3 to 6 weeks to construct its intricate nest. The male chooses a convenient thorn bush and starts stuffing moss and plant fibres between its branches. Later the female helps him and they bring lichen and spiders webs to camouflage the outside of the nest and thousands of animal hairs and feathers to line the inside. The nest must be roomy and steady, for it will have to house 2 broods each of 9 to 12 young... and at nights the adults as well. 24 young are not too many for one summer, for many fledglings are killed by marauders, hailstorms and poisoned insects and many more which do learn to fly do not live long enough to mate. However, feeding 24 raucous young requires a lot of energy. Electronic counters left in Blue Tits' nesting boxes have shown that from 3 in the morning one of the adult birds brings food every 2 minutes and that as the young birds grow the number of feeds per day also grows from 310 to 650. Long-tailed Tits are very active and inhabit bushy commons and heaths and also hedgerows, and woodlands in winter.

The graceful Long-tailed Tit (*Aegithalos caudatus*) is resident almost throughout the whole of Europe except for the extreme North. 6 ins long incl. 3½ ins tail, weighs up to ¼ oz.

The work of cleaning the nest is simplified by the fact that waste material is excreted in little balls with a gelatinus coating which can be removed from the vent immediately and thrown away. Later, the young Azure-winged Magpies will instinctively climb onto the edge of the nest to allow their droppings to fall clear. As long as their bright red gapes show the need for food, the nest is the focus of the parent birds' attention. But after 22 days it is abandoned and the new generation explores the olive groves and pine forests of Southern Spain and Portugal. Some also inhabit orchards and gardens and they breed in small scattered colonies. No Azure-winged Magpie has yet been recorded as having flown northwards over the Pyrenees, whereas our Magpie is resident throughout the whole of Europe, excluding Iceland, Sardinia and Corsica. It is as well adapted to the freezing cold of Finland in winter as to the great heat of Sicily and is perfectly at ease in the territory of the Azure-winged Magpie itself. Both species habitually roam the countryside in small noisy bands, looking for food. The Magpie has a loud voice and utters a series of rapid cries – "chack-chack-chack-chack". – It inhabits open country and farmland, building its domed nest in bushes, hedges or tall trees. The Azure-winged Magpie on the other hand usually builds its open nest in forks of pine trees, or sometimes oak or poplar.

➤ Azure-winged Magpie *(Cyanopica cyanus)*, resident in the Southern half of the Iberian peninsula. 14 ins long, with tail 7½ ins.

➤ "Two-way traffic" in the nest of Bonelli's Warbler *(Phylloscopus bonelli)*: while one parent bird enters bringing food, the other takes out the "diapers". These Warblers (length 5 ins) nest in undergrowth around trees and live in woods where the foliage is dense. They are summer visitors in Central and Southern Europe, making their way back to tropical Africa in autumn. They have been known to wander to Britain and Sweden.

A picture of a Flamingo has been found on the wall of a cave in Spain, painted during the stone age, perhaps as a magic spell to give good luck to the hunter. Ever since then the Flamingo has been hunted recklessly by human beings. At the time of the Roman empire it provided eggs and meat for the common people, tongue and brain for the rich and feathers for their ladies. In the last 100 years alone these flame coloured wading birds have lost 112 of their traditional breeding places, so that today there are only 20 colonies left in the whole world, only 3 of these in Southern Europe, on the western shore of the Caspian Sea south of Baku, in the alluvial soil of Guadalquivir in Southern Spain and at the salt works of Giraud on the Rhone delta. For a whole decade it has been the task of the biological research stations at *Le Sambuc* and *Salin de Badon* to protect and study the *flamants de la Camargue*. In order to understand the colonial existence of the Flamingoes, we need to understand many things; for example, why they should choose the barren land of the saltmarshes on which to live. Also, why they have been observed to breed only 21 times in 45 years and this at quite irregular intervals. We need to know whether breeding depends not only on the weather, but also on the amount of plankton which is available for them to filter out of the slime of the marshland with their unusual beaks, rigid underneath and moveable above. A colony of 6000 Flamingoes will go for 2 or 3 years without reproducing at all, until suddenly during April they all begin at once to clear the leathery plants from the marsh, pound the wet sand and start piling up their mounds of mud with feet and beaks. There is a collective display ceremonial, leading to erotic mass hysteria: among a sea of beating wings the birds pair off and presently a single egg is laid on each little mound of mud. The members of the colony all brood for a month, until the tiny young hatch out. Their first food is a red liquid trickled into their beaks by both parent birds, and their first enemies are the Herring Gull, thunderstorms and noisy jet planes... if 50 per cent of them survive it has been a good year. On their fifth day of life they can already escape from danger by running and swimming. It is doubtful whether each one can find its parents again. The young Flamingoes learn to fly in their third month and for 2 years they are greyish-brown in colour. For the first 6 years of life, i.e. until they are sexually mature, many roam great distances southwards; however ⅓ of the Camargue colony *(see page 3)* stays there during the winter, in spite of the danger of frost.

Flamingoes *(Phoenicopterus ruber)* in the zoo at Basle, which was the first zoo in Europe to provide the right sort of habitat and food for them to keep their rosy plumage and even persuade them to breed. Flamingoes are 50 ins long and walk in a sedate manner.

On the lakes of Central Europe the long legged Herons have become isolated breeders, although it is their custom to breed colonially: for a long time fishermen shot them in large numbers — and when they at last became protected birds there was already a new threat to their survival, the reclamation of marshland. However, in the extreme South East of Europe the Great White Herons still nest peacefully and in reed beds or in trees near water about a hundred pairs are often to be found nesting in a small space. A Heron colony is full of movement and noise and at breeding time there is a majestic display ceremonial. Both parent birds share in bringing up the young, which cannot fly before 2 months. Until then the adults have to catch their food, up to 260 lbs of it: fish, and some frogs, snakes, water voles and mice. The Herons glide silently up to their victims and seize them very suddenly with their darting bayonet-like beaks. The cervical vertebrae are set in a most unusual way, in order to facilitate the jerky, precise neck movements of the Herons — which are unable to groom and oil their plumage like other birds. Instead their plumage is kept waterproof and neat by a greasy powder dispensed by special feathers. The Heron inhabits lakes, sea shores and water meadows, where it will stand for long periods without moving. It also perches in trees. It flies powerfully, with slowly beating wings.

The Heron *(Ardea cinerea)* is 36 ins long and weighs 1¾ lbs and is the largest European species of Heron.

Cranes are migrant birds which return to Europe in the spring in symmetrical formation via the Bosphorus. On migration they will alight on marshland to refresh themselves with young shoots and plants and then continue until they reach their traditional breeding grounds in the Balkans and on Northern plains. Their display ceremonial is most impressive: the male Crane utters its ringing trumpet call to attract the female; when he raises his wings, she retreats, and he follows her to dance round her with a springy step, and sudden high leaps. Then he pulls up tufts of grass and throws them into the air, catching them again playfully. After this symbolic offering of nest material the pair soon build their huge nest and the female lays her 2 eggs. The russet young hatch after a month. They immediately learn to swim, but cannot fly until shortly before their migration to the Nile. The Demoiselle Crane is much smaller than the Crane and has a large tuft of white feathers curling downwards behind each eye. During its slow and powerful flight, its neck and legs are extended. It is a sociable bird except during the breeding season, and builds its nest on dry ground, while the Crane breeds on marshland, reed beds and swamps.

The blue-grey Demoiselle Crane (*Anthropoides virgo*) ➤ breeds only in Roumania.

Pair of Spoonbills *(Platalea leucorodia)* and offspring. This species is related to the Ibis and breeds colonially.

Every spring in the reed beds at the edges of the shallow Neusiedler Lake on the border of Austria and Hungary there are great crowds of Coots, Ducks and wild Geese, Herons and Bitterns, Spoonbills and Ibises, which have all returned to one of the last remaining bird sanctuaries in Central Europe. On these 185 square miles of water, never deeper than 70 ins, all find enough food for themselves and their brood, each one hunting and fishing according to the shape and size of its beak. The Purple Heron struts through the water meadows, the Bittern lies in wait for frogs in the shade of the reeds, while the Great White Heron with its pure white feathers does not need to avoid the hot sun and fishes far out in the open waters using its beak like a harpoon. The Spoonbills, however, wander over long stretches in small groups regimented in rows in the shallow water, continually filtering the slime at the bottom through their bills. The small fry and larvae which they discover there are immediately swallowed and later regurgitated from the stomach to feed the young. These hatch after 24 days in colonies where the nests of the noisy Herons and Spoonbills, which clatter their bills when excited, are built very close together. Spoonbills, which are 34 ins long, often carry reed stalks 2 yards long to the large nest in a reed-bed, bush or tree.

White Pelicans *(Pelecanus onocrotalus)* still breed in Bulgaria and Roumania, and sometimes fly to Greece, Italy and Spain.

Today pelicans still glide like white barques over the wide Danube delta, with its 650,000 acres of reed beds. Pelicans, with swans, are the heaviest flying birds and weigh from about 20 lbs up to nearly 40 lbs; nevertheless they float on the water as lightly as corks. Their bones, plumage and even the subcutaneous connective tissues contain so much air that the birds ride easily on top of the waves. They prefer shallow waters and there make well-planned attacks on the fish: in a group they form a semi-circle and swim in close formation towards the shore. Steering with their strong webbed feet they drive the fish in front of them, until they are all in the middle of the semi-circle and each bird can fish to its heart's content... with its enormous lower jaw serving as a game bag. At breeding time hormone action produces a rosy hue on the pelicans' plumage. The 2 to 4 brown young, which hatch after six weeks, must for a long time be fed with regurgitated fish. Pelicans, which have a daily consumption of 4 lbs of fish per adult, mainly frequent inland waters, and are finding it hard to survive in Europe; even on the lakes of Africa, huge colonies have shrunk to a few thousand birds. However, off the coast of Peru some hundreds of thousands of Alcatraz or Peruvian Pelicans still breed every year.

The White Stork makes a clapping noise with its beak when it is pleased, but the Black Stork claps as a sign of anger. When the White Stork is angry it hisses, but this noise is a sign of great tenderness in the Black Stork. But they are not likely to meet each other, for the Black Stork nests secretly and in isolation in the marshy forests of Poland, Roumania and Russia, while the more sociable White Storks have always preferred breeding in communities. Today those villages and small towns where every second house bears a stork's nest are only to be found in Eastern Europe and occasionally on the Iberian peninsular. These birds have been frightened away from Central and Northern Europe by industrialisation, high tension cables and the reclamation of marshland. The few storks which arrive there frequently do not return the following year, since in migrating westwards via Gibraltar and the edge of the Sahara they often fall victim to shooting or to the toxic chemicals used to combat locusts, their main food. The hosts of storks which go eastwards take a leisurely three to four months for their flight over the Bosphorus and down the Nile to South Africa.

← Young Black Storks (*Ciconia nigra*).

→ A pair of White Storks (*Ciconia ciconia*), which may live together for many years, pairing each spring.

It was something like 300 million years ago that lizards, which had developed rudimentary wings, first tried to glide and after a long process of adaptation broke away from the reptiles to become primitive birds. The only reminders of that clumsy beginning are fossils and the horny scaled feet of today's birds. We see the magnificent results of the evolution: the flapping flight of the song bird, the darting circles of the swallow, the calm undulating flight of the heron, all these are technical marvels, which counteract the force of gravity. Birds' wings are highly complex structures: each wing-quill is made up of nearly a million barbs and barbules, the barbules on the upper side of one barb being caught in the barbules underneath the next. When the wings are beaten backwards and spread out into a fan, the feathers press down on the air and lift the bird's body upwards. And with each lifting of the wing the quills turn, in order to let air flow through the space in between and to make it easier to move against the resistance of the air. The large birds that glide against the wind owe their weightless soaring to the powerful breast muscles which enable them to hold their mighty wings so long outspread against the uprising warm air currents.

Above: The Common Tern in flight *(Sterna hirundo)*, wing span 31 ins. *Middle:* Avocet *(Recurvirostra avosetta)*, wing span 30 ins. *Below:* Curlew *(Numenius arquata)*, wing span 42 ins.

Although every bird preens its plumage carefully every day, even the finest plumage becomes worn by wind and weather, thorns and branches and is renewed by means of the annual moult. In most species, the new feathers grow gradually, so that the equilibrium is maintained and the bird remains capable of flight all the time. This also ensures that migrant birds have new wing surfaces ready for their long flight southwards. Not until the Dane, Hans Christian Mortensen, began ringing birds systematically in 1899, did individual birds, picked out from the vast numbers of migrants, indicate — if they were found again — the routes taken by that particular species. Since then ornithologists from all continents have ringed well over 10 million birds. Now we know the distances travelled by different birds, by large flocks and by isolated inexperienced young birds. We believe that birds flying by day take their direction from the sun, and those flying by night from the stars. But what drives them on and what brings them back? Partial migrants are driven by cold and lack of food — but the real migratory birds leave in late summer when insect food is still plentiful. They leave before the days get shorter and return when they get longer, because they need a maximum of insulation. Through the bird's eye, sunlight stimulates the endocrine glands which again release the stores of energy so vital to these highly energetic creatures.

Above: Squacco Heron *(Ardeola ralloides)*, wing span 29 ins. *Middle:* Little Egret *(Egretta garzetta)*. Wing span 71 ins. *Below:* Night Heron *(Nycticorax nycticorax)*, wing span 42 ins. — All 3 are vagrants in Britain.

Display of Ruffs *(Philomachus pugnax)*. They pair in the marshlands of northern Europe.

The ceremonial display of the Ruffs takes place on the banks of dykes in the early morning and it is difficult to believe that the colourful males have spent the winter with plumage resembling that of the smaller female or reeve. With the coming of spring their gonads become active and their ruffs and ear-tufts begin to sprout... and a great variety of colours is produced. All of the male Ruffs (1 ft tall) are different: white, chestnut, purple and black spots, rings and lines decorate the dancing males. A thousand nerve-ends convey increased stimuli to the numerous tiny muscles which ruffle up the feathers. The participants arrive at the traditional ground for the tournament and the spectacle begins: dancing like marionettes, shuddering right up to the tips of their feathers, they stand side by side, apparently only concerned with themselves. But if one comes too close to another the challenged bird springs up to defend itself, beating its wings. This starts them all off and suddenly the feathers fly around the wildly fighting rivals... until all of a sudden one goes down on its knees, bows its head and goes rigid. They all hold their breath and the plainer looking reeves, which up to now have been merely interested spectators, come forward to break the spell.

Display positions of the Great Bustard *(Otis tarda)*, which inhabits open treeless plains of eastern Europe.

Few ornithologists can claim to have seen the Great Bustard displaying — and then only from a distance — for this giant relative of the rails and plovers is very wary. With its wings spread out to over 6 ft it flies in low to land on the mist-covered plain. Then it begins its curious dance, stamp — stamp — stamp, heavy and trampling. With its thick neck inflated and its whitish bristles ruffled up, the lone dancer circles, nodding its head, without uttering a single sound. As if in a trance it carries out this ceremony without a partner — and suddenly, in growing ecstasy, its grey-brown plumage is literally inverted and the clumsy giant becomes a dazzling white feathered blossom. Here and there, about 100 yards away, others can be seen shining, attracting the hens, which are smaller and outnumber the males. They are very shy birds and the males do not collect a harem, but go on their way after pairing with different females, while the latter incubate their 2 to 3 olive green eggs in fields of roots and corn. Sometimes a male may attach itself to a group of mothers with families, as a guard. Young males join together in flocks, and in their fifth year, weighing over 20 lbs and having at last reached sexual maturity, they pair for the first time. They are about 40 ins in height, the female about 30 ins.

Eiders *(Somateria mollissima)*. The most southerly breeding grounds lie on the Northern European coasts.

During pairing the drakes are more colourful than when they are in eclipse, becoming white on top, with black bellies and a black crown. The ducks are black and mottled brown so as to be inconspicuous while brooding. Eiders mate from April onwards, although they have chosen a partner months before. They are the biggest sea ducks and greatly prefer salt water, landing occasionally on inland waters during their winter migration. They have difficulty in taking off again: they tread water and beat their wings, trying to get enough air under them to lift themselves up noisily into flight. On the other hand diving is easy for them. While they are peacefully searching for shell fish and crustaceans, their wings are kept in special pouches — but if a Sea Eagle circles overhead, they dive quickly and get away under water with the help of their wings. However they do not actually get wet, nor even cold in the frozen North, because of their warm eiderdown, their bulky armour of feathers which contains so much air. Also their plumage is lubricated daily with the water-repellent secretion of the coccygeal gland.

Mallards *(Anas platyrhynchos)*. These wild ducks are to be found all over Europe on many kinds of waters.

Mallards mate on the water, the female (or duck) being largely submerged during the pairing. The 8 to 14 eggs are laid in a cup-shaped nest among bushes or reeds. Incubation takes about 4 weeks. Whenever the duck leaves the nest it lays a fine, felt-like feather cover over the eggs. The colourful plumage of the drake — the shining green head, white neck and purple-brown breast — lasts from late autumn until spring. In summer it becomes a darker version of the duck and can only be distinguished from the female by its darker upper-parts and yellow bill.

The second day after hatching the young can already swim, being the earliest ducklings to mature. They simply lap their food up from the surface of the water but soon they imitate their mother and "duck", heads down, tails up, to forage for duckweed and other titbits. Their ridged bill is constructed so that all superfluous water runs out again at the sides. It is some time before the young can fly; however, once they can, they lift themselves upwards perpendicularly by powerful wing beats without any starting run. Their flight is direct and rapid.

Brooding Avocet *(Recurvirostra avosetta)*. On hot days it spreads out its wings to shield the clutch from the sun.

Avocets breed in colonies near shallow water and on sandy river deltas in scrub, for example in the Austro-Hungarian lakeland. They also breed in the Camargue, in Holland and Denmark, by the Black Sea and in Portugal; and there has been a colony in Suffolk since 1947. Before pairing they stand stiffly in the shallows and, with apparent indifference, draw feather after feather through their scimitar-shaped beak and suddenly spring at each other. As a preliminary to nesting they scoop out several hollows in the sand, choose the best of them and line it with roots and grasses. Both male and female brood; the shifts begin and end ceremonially. The brooding bird gets up, advances politely towards the approaching bird, and both birds bow. Now the relieving partner struts towards the nest and carefully turns over each egg, before settling itself. The other bird then leaves with measured tread, walking quickly and gracefully on its long, blue-grey legs. It may stand up to its belly in the murky water, scooping sideways into the silt, its strange beak filtering out plankton and all kinds of larvae. It rarely ventures into deeper water by swimming.

Black-throated Diver *(Gavia arctica)* on the nest. Both sexes are similarly patterned in black and white.

Black-throated Divers are maritime birds which spend the winter on sea coasts, but in the breeding season they are to be found by inland lakes or lochs in Scotland, Scandinavia and North Eastern Europe, where they nest on islets washed by calm waters. During the breeding season their shrill wails, shrieks and barks can be heard through the night. Incubation takes a whole month, with both male and female brooding. The young, covered with grey down, soon follow their parents into the water and begin to practise diving. Adult Divers can remain under water for 90 seconds, reemerging from depths of 5 to 6 yards at a distance of 30 yards. But at the first sign of danger they sink silently and inconspicuously up to their neck, by squeezing their feathers together to push out the air trapped between them, thus adapting their specific weight to that of the water. In winter they rest on rivers and lakes on their way to the coasts of Southern Europe. Once their breeding plumage disappears, they have grey backs and white throats. These changes in the colouring of birds are brought about by the action of hormones on the pigment cells.

Great Crested Grebe *(Podiceps cristatus)* and family. It is 19 ins long and the largest of the Grebes.

It is unusual for young birds to have conspicuous stripes instead of a colouring which blends with the environment — but in the case of the Great Crested Grebe both colouring and instinctive behaviour ensure the survival of the species. The female never loses sight of the striped young when they are swimming on lakes or reservoirs, and when danger approaches the young disappear into the plumage of the parents. At about 6 weeks they are able to dive under water. On hatching, the tiny bundles of feathers climb straight on to their parent's back, for after 4 weeks' incubation the water has reached a dangerously high level in the floating nest. The young are carried everywhere pick-a-back or hidden under the wings. The parents feed them with small fry and insect larvae and they grow fast. — One couple will stay together for many years and in spring the display consists of the two birds facing each other and wagging their heads. The male, with his crest and ear tufts raised, swims up to the female with wings unfurled, dives and reappears near her with a water plant in his beak. She reciprocates and then they dance with their necks entwined.

Black-necked Grebe *(Podiceps nigricollis)* by the nest. It has golden ear tufts at breeding time.

All the grebes lay their eggs at two-day intervals and start brooding as soon as the first egg is laid. Thus the first-born is already being carried on its parent's back when the rest are still in the eggs. By the time the youngest is hatched, the eldest will have been swimming for some time and no longer needs to climb onto the "raft" over its mother's outstretched foot. The legs of the grebes are set well towards the rump and act as powerful rudders, being lobed instead of webbed. Because of the position of their legs they avoid the shore, for they can only totter on land. The edge of the reeds affords enough protection and their enemies have become scarce on the lakes of Central Europe. But where grebes are present in larger numbers, fishermen complain about their effect on the fry. To fishermen a lake is only a reservoir of fish; they do not see it as a biological community. There is a balance in nature between harm and good: what the water birds extract from the water they also return as manure, and in moving about, they are spreading small living creatures and water plant seeds which cling to their plumage from one water to another.

The male and female Stone Curlews have an exact timetable for brooding: from 4 in the afternoon to 3 the next morning the mother sits on the two grey-brown eggs and the male sits on them during the hottest part of the day. On the 27th day of incubation when one bird relieves the other they find that one of the chicks has hatched. It is not long before the second hatches, and by the evening the nesting hollow will already have been abandoned. They hide in crops or patches of heather, with their young, making forays to catch all kinds of small invertebrate animals. For another two weeks the young remain in shelter, but then they too take part in the hunting trips in the evening. With their large yellow eyes Stone Curlews are truly crepuscular and only after sunset do their voices rise from the low note of the oboe to a shrill flute-like sound. Their short bills are black and yellow in colour, and the birds look large and awkward, their roundish heads distinguishing them from all the others waders.

Stone Curlews *(Burhinus oedicnemus)* have become rare in Central Europe, though they still nest in South-eastern England. Plumage streaked with brown and white. Adult length 16 ins, wing span 33 ins.

While the female tern is brooding the male often brings her a small fish. This is a reminder of the display ceremonial, when the male approaches the female with a silvery fish which he offers her again and again, until, after hesitating for hours, she finally accepts. Thus she symbolically accepts the male. The pair then find a small place in the scattered breeding colony — usually on a beach or the shore of a lake: Little Terns do not nest close to each other like their bigger relatives. When the young hatch after three weeks' incubation their first meal consists of tiny crustaceans and larvae and later they receive the smallest possible fish. On hatching they are the size of a thumb. They are precocious birds; they can already walk at two days old, so that their parents have to keep a sharp look-out to warn them of the attacks of seagulls and to protect them against cloudbursts and sandstorms. Adult Little Terns have yellow legs and yellow bills tipped with black, with a grey crown in winter and black in summer.

Little Terns *(Sterna albifrons)* — 8 ins long, wing span 20 ins — are the smallest of the terns and breed in colonies on isolated sandy coasts all round Europe except in Scandinavia and Sicily.

The gulls, with their keen eyesight, get their food from the sea's surface, but the creatures which live in the depths are safe from them. Though they do not dive, Herring Gulls are useful as scavengers, for no dead fish escapes them. They immediately devour anything left behind on the beach by the tide or revealed in the shallows when the tide goes out, and what is thrown overboard by fishermen and ships' cooks. The gulls' raucous calls are in fact a language full of variety, with each different note having its own particular meaning for other birds of the same species. They hear the call of other gulls miles away and follow it until they find themselves flying in the wake of a ship or on the route taken by fish living at or near the surface. Immature gulls with their grey-brown markings are to be found among the swarms of silvery grey and snow-white birds. These gulls will not attain sexual maturity and the plumage of the adult until the age of three years.

Herring Gull *(Larus argentatus)*, length 24 ins, wing span 5 ft. The commonest of all European coastal gulls, nesting all round the coasts even as far as Iceland. At breeding time they rob the nests of smaller sea birds and in most countries their eggs are not protected by law.

Young Gannets, which are fed for months by the shining white parent birds on the most delicious fish, suddenly find in September that they have to fend for themselves. The adults visit less and less the seaweed nest perched on a ledge on some steep cliff, and eventually stop feeding the young altogether. There is nothing left for the hungry young Gannets but to plunge exhausted into the sea. For days they swim around in thousands – an easy prey for sharks – with no apparent object, and would starve surrounded by an abundance of food if it were not for the fact that they can live off their own store of fat. Finally they begin to try their wings out and gradually they learn to fly and then to dive, plunging headlong into the water in the manner of an adult Gannet. From now on they catch their own fish from shoals which pass within their reach – and like all sea birds they drink all the sea water they need, the salt being discharged via the nasal glands.

Young Gannet *(Sula bassana)*, *see also pages* 5 *and* 30. Gannets lay only one egg weighing 3½ ozs. The dark-skinned young Gannet hatches after 43 days' incubation and soon grows a covering of light down, which is replaced after 9 weeks by the speckled plumage of the fledgling.

Along the southern coasts of Spain: Barracudas or Spet *(Sphyraenidae)* harrying a shoal of Gold Lines *(Box salpa)*.

For sea birds there are only two kinds of fish: those which can be preyed on, and those which cannot. Gourmets distinguish the particularly choice fishes from the edible but less choice. But anatomists and palaeontologists divide the fishes into two major groups according to their degree of evolution: the ancient elasmobranchs such as sharks and rays, which have a cartilaginous rather than a bony skeleton, and the bony fishes or teleosts, which arose some millions of years later. Biologists study both the pelagic life in the upper layers of the sea, and the benthic life of the sea floor. Ichthyologists have so far named some 25,000 species of fish, and the list is still growing, for at least two thirds of all vertebrates live in water, that vast, primaeval medium. Few fishes are truly cosmopolitan, most live within the limits of a particular range of water temperature, pressure, salinity and illumination. Many species stay all their lives in the waters of the Continental shelf, where fish movements are mainly concerned with avoiding adverse changes in the environment. But during the breeding season some fishes seem — by biochemical changes in their own body — to be drawn to the very places which they normally shun: some travel to shallow beaches, others descend to the dark depths, some seek the richer oxygen of surface waters, and a few enter the less mineralized inland streams, each species finding by instinct the biochemical environment most suitable for the development of their offspring.

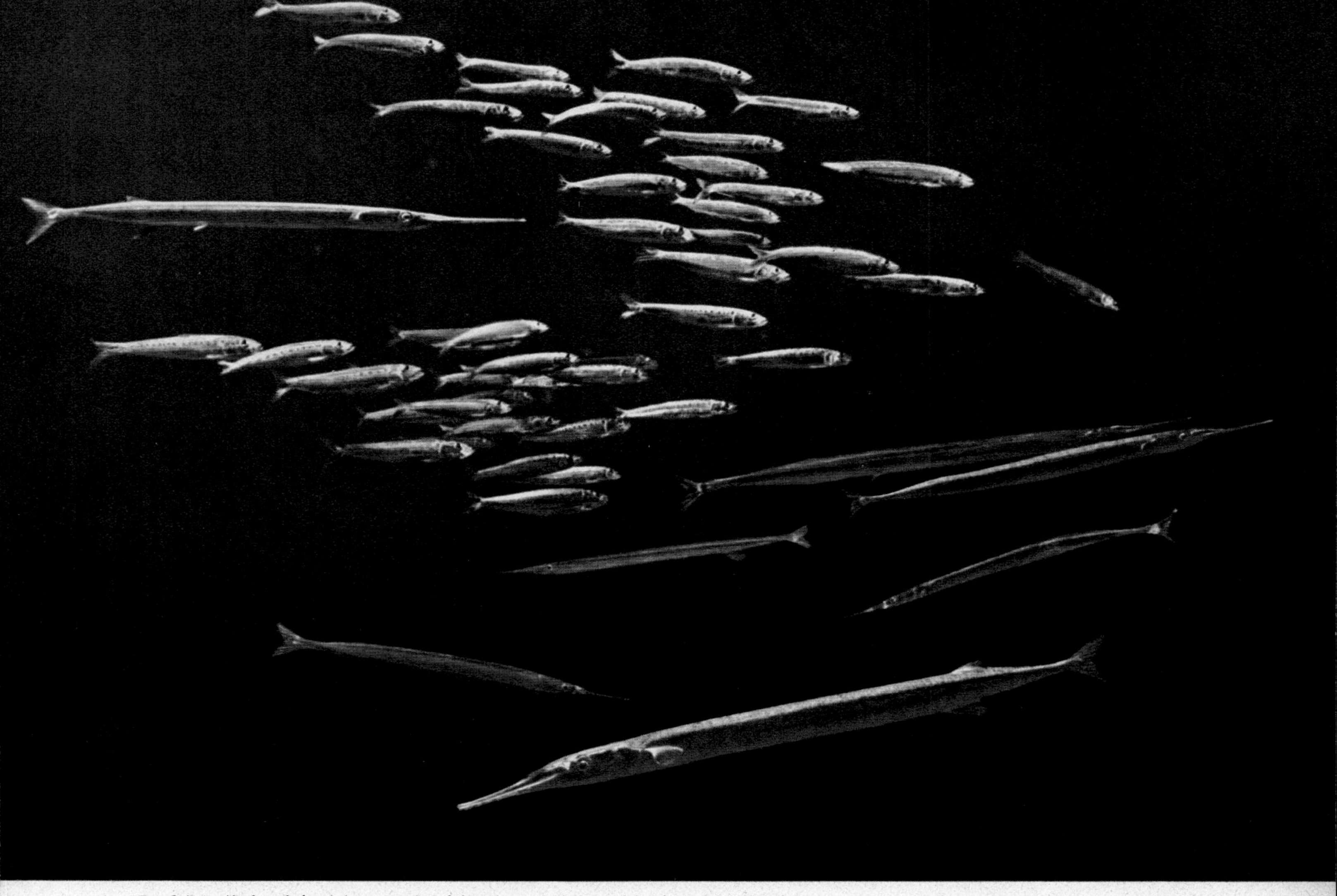

Garfishes *(Belone belone)* in pursuit of Sardines *(Clupea pilchardus)*. The Garfish can attain 3 ft in length. Its bones are green.

Gregarious fish are perhaps the most hunted of all the vertebrates. Only amazing fecundity preserves the species, provided that they deposit their millions of eggs and sperm in precisely the same place and at the same time. Thus the fish are obliged to travel in shoals, although such assemblies of shining bodies attract predators. In the face of danger there is only one alternative — flight. The large, round eyes of fishes inspect their immediate surrounding, whether asleep or awake; they can also be focussed for longer distances. The sense of balance in fishes is located in the "ear". Solitary fishes need only to hear the approach of enemies and so make their escape. But social fishes must also be able to warn their own kind, and this is achieved by a variety of grunts, hisses, and groans, imperceptible to man, but none the less belying the old conception of the "silent world". Some truly mute species secrete an "alarm substance" when wounded or attacked, which their companions smell, even though highly dilute. Fish do not only smell and taste by means of the nasal fossae, lips and palate, but also with the olfactory and gustatory organs located in the lateral lines along the flanks. Thus fishes can sense chemical changes in the water as well as feel currents and eddies.

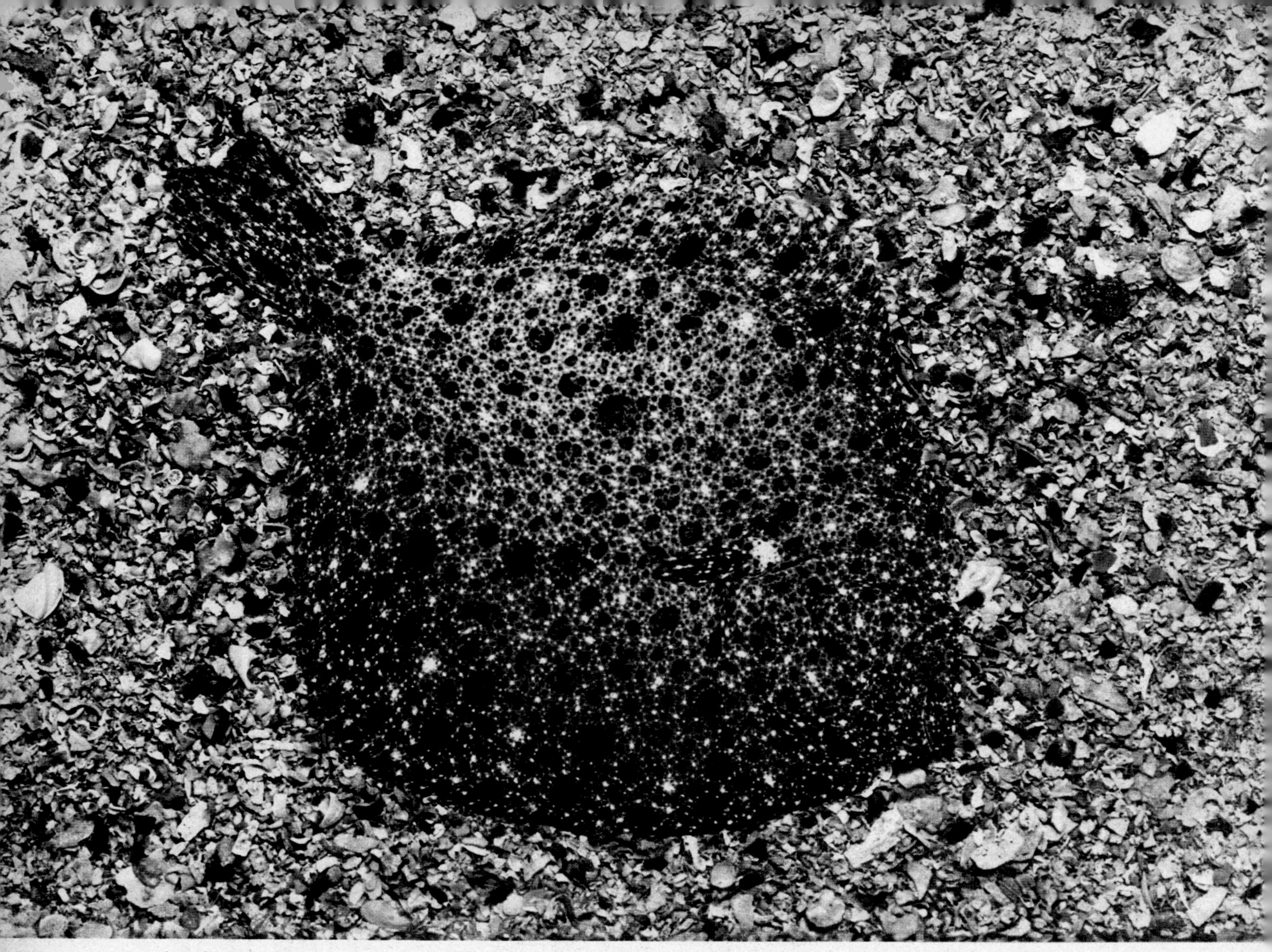

A well-camouflaged Turbot *(Scophthalmus maximus)*. It does not breed until it is 5 years old, and can exceed 3 ft in length.

Not all fishes are stream-lined, nor do they all swim in the well-lit upper layers. Flat fishes, such as the Turbot, the Dab, the Sole or the Plaice, live on the bottom. There they rest, ever watchful, camouflaged either through their ability to change colour by means of their chromatophores (colour cells), or by burying themselves in the sand or mud. With their mobile eyes, they watch everything that passes by. When the Turbot sees its prey, it suddenly rushes up on it or seizes it as it passes overhead. The flat fishes are unique amongst the vertebrates in one respect. From their eggs emerge transparent larvae which show perfect bilateral symmetry, the back uppermost, the belly below, and the eyes placed on either side of the head. But after about a fortnight one of the eyes is moved to the other side by an asymmetrical development of the skull. Internal organs and skeleton little by little become laterally compressed to produce a flat, asymmetrical fish. According to Juvenal, Turbot was esteemed as a delicacy even in Roman times. Tastes have not changed much since then, for in Europe alone some 8,000 tons are eaten each year.

A well-disguised trap: the Angler fish *(Lophius piscatorius)*, almost invisible. It reaches 6 ft in length.

For fishes passing overhead, this looks like a peaceful stretch of sea-bed, but their fate is nearer than they think! The Angler lies motionless, the fringed border ornamenting the immense mouth and sides of the body softening its outline. When the Angler sights passing fishes it raises the first ray of the dorsal fin, which has at its tip a tag of skin. Briskly jerking the tag to and fro, the Angler attracts their attention and they swim towards the moving lure to investigate its suitability for food. Just as they reach it the Angler with a sudden upward heave of its huge head snaps open the enormous mouth to suck in the inquisitive prey. Once closed on a fish the backwardly directed hinged teeth of the jaws ensure no return for the unfortunate victim. "Nothing but head and tail" says an ancient description of this fish, to which one might add "and stomach", for a full-grown Angler can swallow prey as large as rays and young sharks. It travels long distances only in April when it returns to the spawning grounds, which are in under-water valleys between 1,200 and 6,000 ft below the surface. The spawn of the Angler fish is remarkable in that a single female forms a floating sheet of a million or more eggs embedded in a jelly-like substance 20 or 30 ft long and 2 or 3 ft wide.

This odd little fish, discovered in 1884 in the straits of Messina, was named *Eretmophorus kleinenbergi* in 1889.

What enables a fish to swim? Mainly the swim-bladder which operates as a regulating hydrostatic organ so that the fish can float at different depths by adapting itself to the pressure of the water. The more stream-lined the body the more easily it can glide through the water with the aid of the fins. The caudal fin is used for propulsion and for steering, the pectoral and ventral fins control ascent and descent, and a dorsal and an anal fin maintain the fish in a vertical plane. This is the normal condition in most fishes. But there are fishes brought to the surface from the depths which have fins modified into tactile filaments or fishing rods or supports for spines. Such modifications are usually rare in pelagic fishes. Therefore one can imagine the surprise of Professor Kleinenberg of Messina University at the small (3 ins) unknown fish which he caught at the surface in April 1884. Instead of the ventral fins, this fish had five filaments, three of which terminated in leaf-like blades like little "oars". When two further specimens were caught, this new fish was named *Eretmophorus kleinenbergi* — Kleinenberg's little oarsman. Recently, the Italian zoologist D'Ancona has expressed the opinion that it may be a stage in the development of *Haloporphyrus lepidion*.

This young Flying Fish *(Exocoetus volitans)*, one of the rare Mediterranean flying fishes, grows to 20 ins.

In the flying fishes the pectoral fins are so enlarged that they form wings, although these cannot be flapped. When chased by a predator (a tunny or a dolphin), or when frightened by an approaching boat, these blue-coloured flying fishes shoot like arrows out of the water and can travel up to 300 yards through the air (being airborne for about 10–12 seconds) before plunging into the crest of a wave. Confronted with further danger, the frightened fish regains the surface as quickly as it can, clearing the water with its folded "wings". Then, spreading its pectoral fins and beating its tail violently, it rises some feet above the surface, richochetting successively like a flat pebble. From the rails of a ship, this unforgettable sight is merely a diversion, like the frolics of the dolphins, but for the fishes it is an anguished attempt to escape death, at the interface between the two elements. In this case, departure from the familiar liquid environment is successful, but this is not always so. Thus, in spite of its immense pectoral fins, the Sea Robin *(Dactylopterus)* cannot really take to the air, its body being far too heavy. The colourful pectorals, being suddenly spread out, are used instead to frighten away enemies.

The St. Peter's fish or John Dory (*Zeus faber*) amongst sea-fans which may reach about 1 foot high.

Why is there a black spot on the flanks of these fishes? Legend has it that this is St. Peter's thumb mark, but for those short of money there is unfortunately little chance that, like the apostle, they will find the reputed coins inside the fishes' gullet. In northern Germany the St. Peter's fish is known as the "King of Herrings" because of its habit of following the young herring shoals which, it is believed, are its flock and on which it undoubtedly feeds, for it is a fish-eater stalking its prey. This curious "shepherd" sometimes reaches nearly 2 ft in length. In the past, it was envied by fishermen who would have liked to have learnt from it the movements of the herring shoals. Only recently have men succeeded in tracing these movements by laboriously marking tens of thousands of fishes with little plastic tags or information contained in tiny tubes. Those herrings which are subsequently recaptured help to indicate the paths of these mysterious migrations, so that nowadays fishermen can make use of scientifically planned programmes, both along the coasts and in the open sea. Thus their catches have become more regular and the whole economy benefits. Each year millions of herrings, pickled in brine or smoked, are eaten as the poor man's meat.

A Remora *(Echeneis remora)* attached to the back of a young Stingray *(Trygon pastinaca)*.

↟ The sucker of the Remora.

A terror of the seas, the Stingray even frightens sharks because its long tail is armed with a venomous sting capable of inflicting fatal wounds. But the Remora is not unduly troubled by this. As a means of transport it regularly uses some of the most dangerous animals in the sea, Blue Sharks, Sawfishes, Swordfishes, knowing exactly where to attach its sucker in order to avoid their weapons. Its attachment organ is something of a technical miracle. The first dorsal fin forms on top of the head a suction pad of movable lamellae which can produce such a powerful vacuum that the remora cannot be removed forcibly. It only detaches itself when it is hungry, catching the "crumbs" which fall from its master's table. If it ever misses the boat, so to speak, it can quite well manage on its own, but on condition that it takes into account the change in perspective: as a furtive passenger lying on its back it sees the world upside down, but forced to adopt the more normal mode of life it must readjust to seeing things the right way up.

Two Moray Eels *(Muraena helena)* and the Conger Eel *(Conger conger)*, the latter often reaches 9 ft in length.

According to Pliny, the shout "*Ad muraenas!*" was the order given by Vedius Pollio when commanding that his recalcitrant slaves be thrown into a fish pond writhing with eels. Nobody curses Vedius Pollio, but the Moray Eels have been feared ever since, although the "murderous" eels were themselves held prisoner in Roman fish ponds in their thousands solely for the enjoyment of gourmets. Nowadays, skindivers "throw themselves to the eels", but they know how to harpoon these creatures, being careful to avoid their terrible bite and the venom produced by their palatal glands.

But, more than the Moray Eel, the underwater hunter fears the adult Conger Eel. These giant eels, which can weigh over a hundred pounds, became famous some 70 years ago when the French ichthyologist Yves Delage showed that the little *Leptocephalus*, hitherto thought to be a distinct species, was in fact the larvae of the Conger Eel. Little by little the mystery of the freshwater eels has been unravelled, including their amazing 3,000 mile migration as larvae across the Atlantic from the breeding grounds in the Sargasso Sea area. The breeding grounds of the Conger Eel, however, are still not known. It must be somewhere in the ocean depths, since when sexually mature they become deepwater fishes. Thereafter they can withstand immense pressure but lose the ability to regain the upper layers.

Attaining 6 ft, the Tope *(Galeus canis)* occurs in all coastal areas. It needs some 40 lbs of food a day.

Requin, the French word for shark, is thought to be derived from the word "requiem", implying that when a shark approaches, one's only hope is in an afterlife! But nowadays underwater films have popularized the idea that groups of sharks can be dispersed by shouting or beating a stick. The several families of sharks, together with the rays and chimaeras, are the last survivors of the ancient cartilagenous plagiostome fishes. The group contains giants over 60 feet in length, and dwarfs of only 3 feet, insatiable predators, and harmless plankton feeders. Some sharks, the most ferocious, can communicate with each other by means of short-frequency waves, and that explains why they all come surging round when a quarry has been sighted or a whale has been wounded. These marine scavengers are especially found round floating corpses, indicating that a swimmer is more exposed to attack than a diver. Sharks appear to think that all floating objects are theirs by right, so that one should never swim far from the shore unless equipped to dive underwater. Sharks abound in the wakes of ships, where they find offal of all kinds. Always questing for food, they cannot survive on carrion alone, and so must continuously swim from the depths to the surface and back again, catching a pelagic fish here or a crustacean or mollusc there, for no carapace is able to withstand their savage jaws.

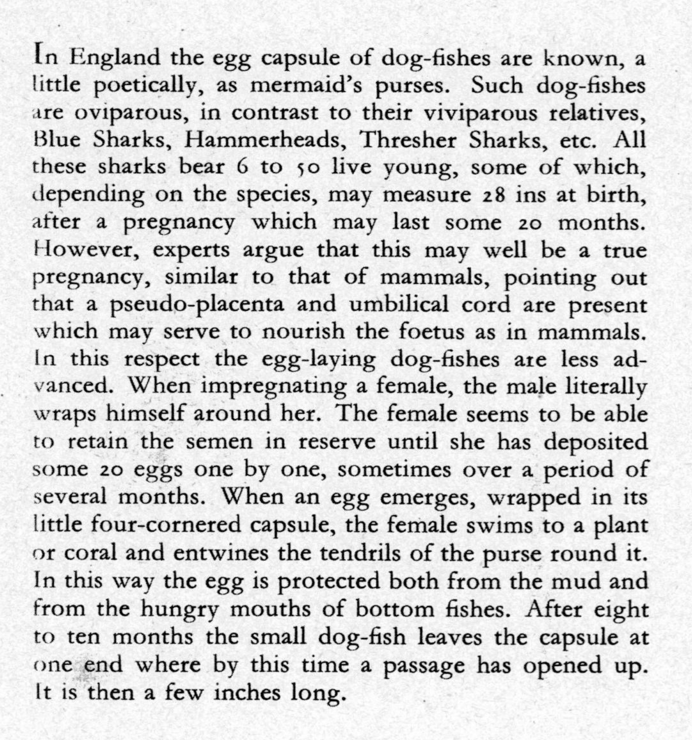

In England the egg capsule of dog-fishes are known, a little poetically, as mermaid's purses. Such dog-fishes are oviparous, in contrast to their viviparous relatives, Blue Sharks, Hammerheads, Thresher Sharks, etc. All these sharks bear 6 to 50 live young, some of which, depending on the species, may measure 28 ins at birth, after a pregnancy which may last some 20 months. However, experts argue that this may well be a true pregnancy, similar to that of mammals, pointing out that a pseudo-placenta and umbilical cord are present which may serve to nourish the foetus as in mammals. In this respect the egg-laying dog-fishes are less advanced. When impregnating a female, the male literally wraps himself around her. The female seems to be able to retain the semen in reserve until she has deposited some 20 eggs one by one, sometimes over a period of several months. When an egg emerges, wrapped in its little four-cornered capsule, the female swims to a plant or coral and entwines the tendrils of the purse round it. In this way the egg is protected both from the mud and from the hungry mouths of bottom fishes. After eight to ten months the small dog-fish leaves the capsule at one end where by this time a passage has opened up. It is then a few inches long.

➤ Egg capsule of the Nursehound *(Scyliorhinus stellaris)* attached to a Gorgonian, one of the Mediterranean corals *(Eunicella verrucosa)*.
➤ "Sea raisins" or cuttlefish eggs *(Sepia officinalis)* see p. 113.
➤ Eggcase of a ray *(Raja)*. The young ray leaves this 4¼ ins four-cornered capsule after several months.

The desire to make a place of one's own is not solely the prerogative of capitalists. Fishes also set themselves up as "landlords" who must defend their territory. An empty oyster shell here provides a home for the small Butterfly Blenny *(Blennius ocellaris)*. ↓ A small pot-hole and a "garden" surrounding it are the domain of this juvenile Scorpion Fish *(Scorpaena scrofa)*. ← A little distance below the surface the skindiver along Mediterranean coasts can amuse himself by trying to spot these rock fish, so beautifully camouflaged amongst the barnacles and sea anemones. But scarcely has the snorkler's movement alarmed the fish, than it as quickly disappears. Harpooning the adult Scorpion Fish (an indispensible ingredient of Provençal bouillabaisse) requires a journey into deeper water. But this "scorpion" is well able to defend itself, and if you meddle with it you are likely to pay for your foolishness. For the anterior rays of the dorsal and ventral fins are furnished with venom, so that even octopuses avoid them.

← These orange beauties are sea anemones *(described on page 105)*. Beside them can be seen the shells of small grey cirripedes — sedentary crustaceans and relatives of the barnacles *(see page 90)*.

Starfishes prey on bivalve molluscs, but the Scallop *(Chlamys opercularis)* is often able to escape by swimming.

On the shell-strewn beach everyone reverts to childhood, for surely here the passion for collecting is born. Each species builds a shell particular to itself, and the different kinds of shell are as diverse as are their different ways of life. There are the sedentary species, in which the lower valve is attached to a rock. Then there are the hidden ones, which bury themselves in the sand, with only their breathing and feeding tube or siphon exposed. Again, some are hermaphrodites, while others are heterosexual. Some species force their way deeply into even the hardest rocks, either by drilling with their shells *(e.g. Pholas)*, or by means of acid secretions. Others send mooring cables, made of filaments from their byssal gland, down to the sea-bed. Some can jump, while the scallops can swim. The latter are quickly alerted to danger, for amongst the numerous tentacles fringing the mantle are many dozens of light-sensitive points. A passing shadow does not necessarily presage the attack of a predator, but should a starfish approach, a chemical excitation indicates its presence. By a sudden muscular contraction closing the valves, the scallop expels jets of water from its mantle cavity to propel itself upwards. Rapidly opening and closing its valves, it swims in zig-zag fashion to a safer place.

The Mediterranean Fileshell *(Lima squamosa)*, a good swimmer, displays here the gills and the pointed foot.

When the Fileshell rests on the bottom with valves apart it lets its cluster of tactile filaments hang outwards, the theory being that it is then mistaken for an anemone. But if, in spite of such pretence, it is attacked, then it is up and away swimming jerkily and surrounded by its undulating crown. Besides these wandering fileshells there are others which are sedentary. They build a little rampart of pebbles and broken shells, binding them with filaments from their byssal glands. This is unusual amongst the lamellibranchs (or bivalve molluscs) whose shell is normally sufficient defense on its own. The shell is built by the two pieces of skin which form the mantle. As a result of glandular secretions and the fixation of certain minerals, the mantle is able to manufacture a shell which continually grows. When the two halves are produced at an equal rate, symmetrical valves are formed, but when only one side of the mantle is active, the surface of the shell develops in the form of a spiral, as in the gasteropods (snails, etc.). All the special features of each particular species' shell, undulations, teeth, rows of spines, colour patterns or pearliness — all these are written into the hereditary constitution of the mollusc.

In giving things and living species a name, men seem to show their mastery over them. But some of the older names are most puzzling. An example is the French word for barnacle, "*anatife*", which can be translated "ducks-nest". In the Middle Ages men claimed that these calcareous shells were the eggs of the Barnacle or Tree Goose; this may be seen from the main door of the church at Moissac. Thus Christian gourmets were able to eat goose at the height of Lent, considering it a kind of bird-fish. This error persisted until the 19th century, when CUVIER replaced it by another error, saying that barnacles were molluscs... In fact the builders of these tiny shells are really small crustaceans. In the early part of their life cycle they are little nauplius larvae which swim in thousands, protected by a dorsal carapace 2/5 ins long and equipped with many horns, spines and appendages. Then, after a complicated metamorphosis, they become sedentary. A flexible muscular peduncle forms which bears the two valves within which shelters this extraordinary crustacean, whose limbs are transformed into cirri. When all is quiet, it opens the shell and waves its cirri to draw fresh water and plankton towards it, but at the least sign of danger, the shells close. For safety's sake these odd crustaceans have become sedentary and yet they sometimes embark on long voyages, attached to flotsam or the hulls of ships. ➢ *Barnacles* (Lepas anatifera), *natural size ½ – ¾ inch.*

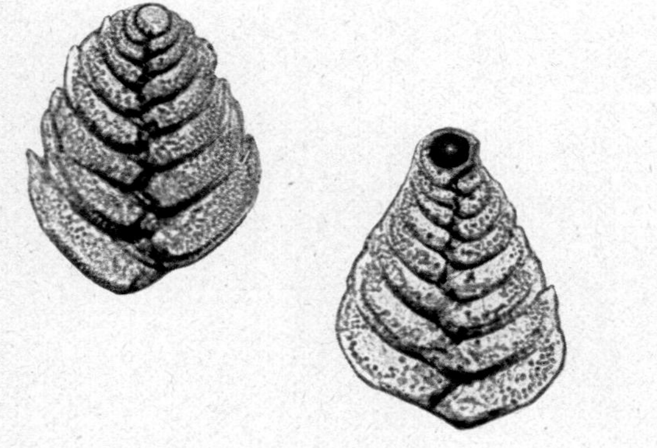

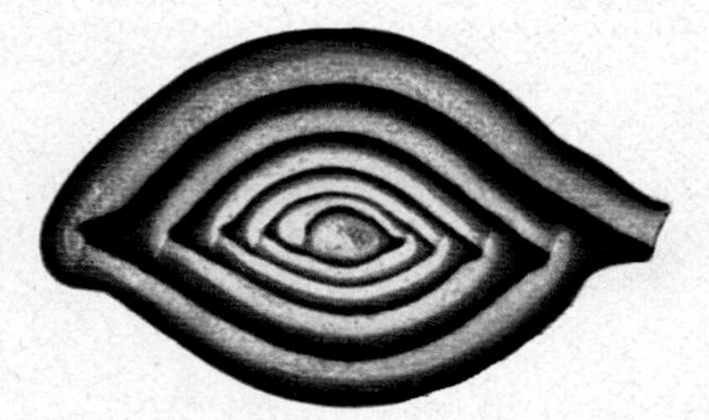

Foraminifera, extracting calcium from sea water, build their tiny shells. In time millions of these shells go to form ocean deposits.

Top: Bolivina robusta Brady, *a very ancient Mediterranean fossil (Bologna). — Middle:* Nonionina depressula, *a present-day species from the North Sea. — Lower:* Spiroloculina limbata, *an ancient fossil foraminifera from the North Sea (Holstein). Magnified 60-100 times.*

Starfishes are perhaps the commonest souvenirs of a sea-side holiday, but few of us stop to wonder how these echinoderms feed and breed. The rather remarkable radial symmetry results from a long metamorphosis. In fact from the spherical starfish eggs emerge small transparent larvae which are symmetrical only in plan view and disport themselves at the surface. They feed on microplankton, grow and, gaining in weight, sink to the bottom to become starfishes. From now on they do not leave the seabed where they move about very slowly. By forcing water into hundreds of little walking feet along each arm, they are able to elongate the arms, stretch them out, and attaching them to the sea-bed, pull themselves along inch by inch. They are however clever hunters, feeding on molluscs and even crabs and injured fish. By means of enzymes, starfishes break down their prey and so ingest them. When attacked, they writhe and will voluntarily sacrifice an arm if it will ward off the enemy. In really desperate situations they are able to divide into five pieces, each surviving part later being able to regenerate into a new star with five arms.

⤙ A starfish *(Echinaster sepositus)* passing over an ascidian *(Cynthia papillosa)*. Initially mobile larvae, the ascidians later become sedentary. They feed on nutritive substances which they draw in and filter from the sea water. Below are two prawns.

➤ In aquariums one can often witness what are otherwise secrets of the deep: here a starfish devours a small Sea Perch.

Although they owe their name to their radial symmetry, the starfishes are by no means all alike. The tree-like ophiurid known as the Gorgon's Head has branched arms which writhe like knotted serpents, but when at rest this delicate lacery seems inappropriate to the name given to this species. On the other hand, such intricacy and beauty has its drawbacks—what a problem it must be to synchronise movement! Better to have fewer arms, say fourteen like *Solaster papposus*. When this cunning predator on shellfishes selects an oyster for attack, it places itself on top of its victim, brings all its suckers into action, and begins to force the two valves of the shell apart. Then it waits patiently, half an hour if necessary, until the oyster, needing to respire or refresh its store of water, opens just a crack. Then quickly the starfish prises the two valves apart, gorges itself on the defenceless oyster, and settles down to digest it.

⤙ The Sun-star *(Solaster papposus)*, which has eight to fourteen arms, occurs in the North Sea and in the Atlantic as far south as Brittany.

⤚ The Gorgon's Head, an ophiurid, found in all warm seas. In the Mediterranean this family is represented by the species *Gorgonocephalus arborescens* and *Astropartus mediterraneus*.

Chrysaora hyoscella, a jellyfish with a bell often 1 ft. across and tentacles 6 ft. long.

Strange creatures jellyfish, fascinating but at the same time a little frightening, transparent but somehow impenetrable. They travel, with rhythmical movements of the swimming bell, subject to little understood stimuli. All oceanographers have dreamt of unravelling the mysteries of jellyfishes, but it was Adalbert von Chamisso, poet as well as zoologist, who resolved the enigma of their method of reproduction. From the egg hatches a ciliated larva which then attaches itself to the sea-bed and becomes a sedentary polyp, a kind of asexual, intermediate form. By transverse divisions, the minute polyp or Scyphistoma (see drawings) comes to resemble a pile of stacked plates, each of which, breaking away one after the other, gives rise to a little medusa, while other polyps detach medusae by budding. Jellyfishes incorporate so much water in their cell tissues that the swimming bell may be composed of as much as 98% water. But this watery body is perfectly capable of defending itself, the tentacles surrounding its rim being heavily armed with batteries of cells each of which at the slightest touch shoots out a paralysing thread sharply tipped like a dart. An invasion of jellyfish on a bathing beach leaves some painful memories!

Sessile stages, or Scyphistoma, formin

The large Mediterranean Sombrero Jellyfish *(Cotylorhiza borbonica)* with its usual swarm of Horse Mackerel.

Whereas *Chrysaora* has four ribbon-like arms which can seize prey from some distance away, *Cotylorhiza* has a large number of small suckers. But what about all these small fishes that swim so imprudently round the Medusa? And how is it that, at the slightest hint of danger, they take refuge under the bell, emerging a moment later alive and unharmed? The key to this mystery lies in a strange form of symbiosis, which raises further questions. Do these little fishes possess some kind of chemical inhibitor which dissuades the jellyfish from discharging its poison? Or does the jellyfish protect them of its own accord for services rendered, for cleaning up inside it or acting as a lure for larger fishes? Thus a predatory fish, eager to pounce on these small fry, does not stop to ask where the little ones are leading him and the chase ends under the skirts of the jellyfish. It has only to graze the tentacles and hundreds of sharp-pointed stinging threads are shot into its skin, paralysing it, the suction pads gripping it, pulling it to pieces and ingesting it. The drama lasts a few minutes, then the jellyfish goes its way, surrounded by its little band of satellites.

edusae by transverse divisions (strobilisation).

The Jellyfish *Cotylorhiza* in motion, its bell open and closed.

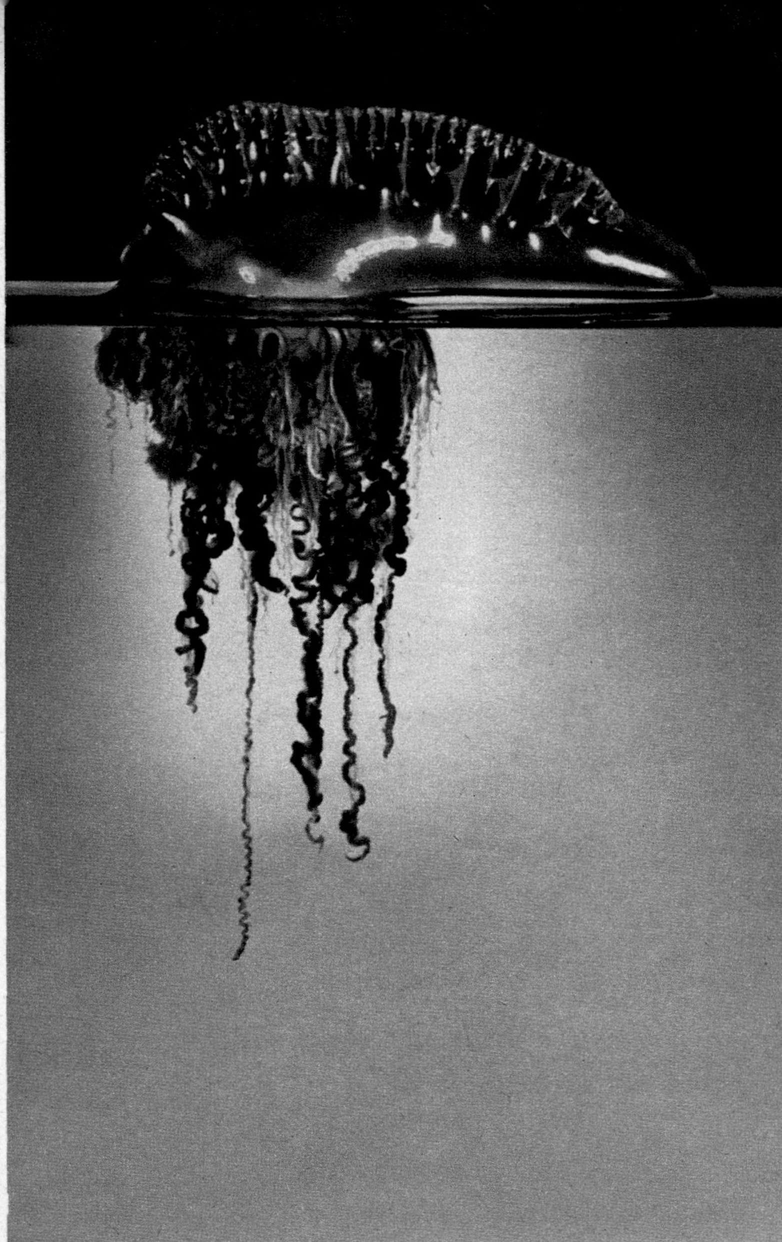

The Portuguese-Man-of-War *(Physalia physalis)*, a complex . . .

In swimming, jellyfishes make the bell or parasol-like disc contract, forcing water backwards and propelling them slowly forward. But how can their mysterious migrations be explained? Do they respond to certain stimuli, to the effects of water currents, light or salinity which lead them to congregate at times in such compact masses that even small boats cannot pass through? This is one of those mysteries to which the oceans have not yet yielded an answer.

Winds and currents across the Atlantic both help to bring the fleets of Portuguese-Men-of-War to British shores from warmer seas. These pinkish-blue jellyfish, which are a complex of many individual polyps, get their name from the fact that they are held up by a gas-filled float topped by a crest which acts as a sail. The long filaments used for fishing reach several yards into the water, ready to discharge stinging-threads into any prey which brushes against them.

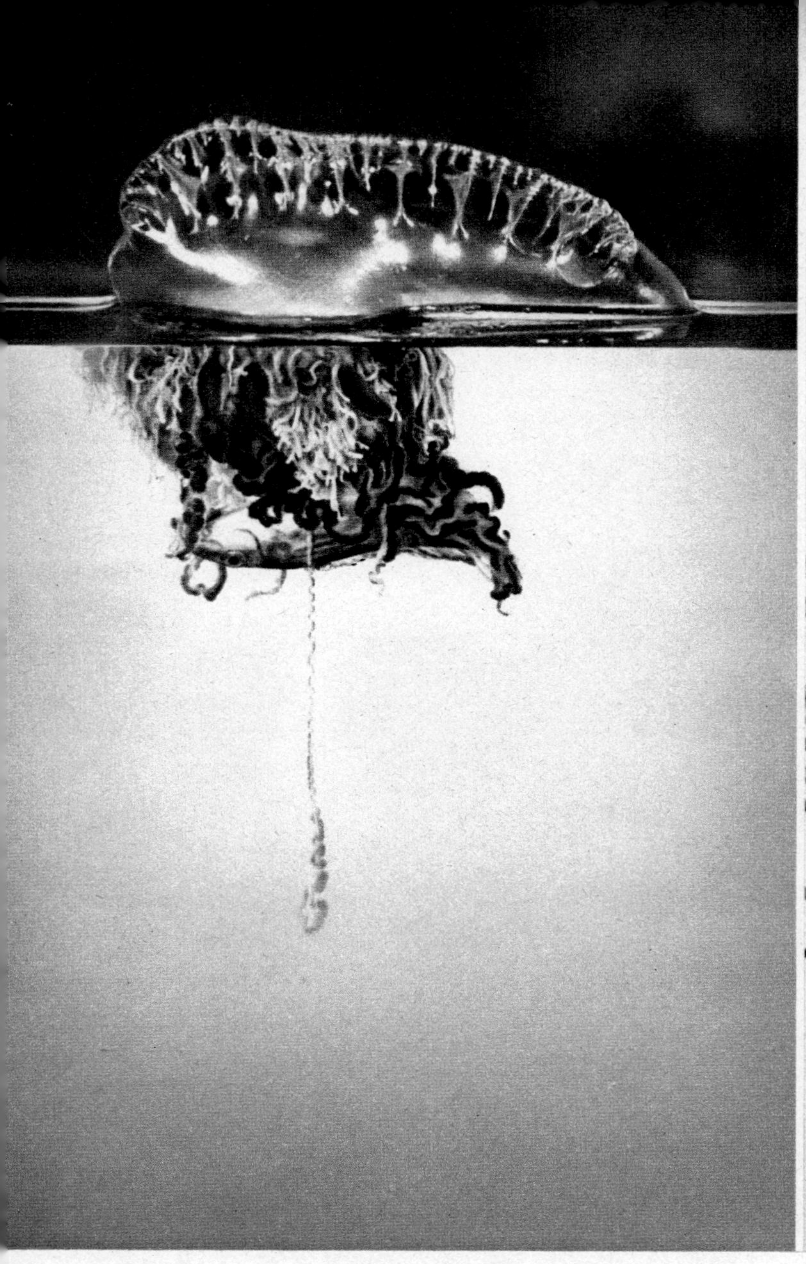

. . . of jellyfishes suspended from a foot long float, devouring a fish.

Halyclystus auricula and *Craterolophus convolvulus*.

The Portuguese-Man-of-War is an astonishing creature, a triumph of collective life. Each member of the assembly, through a division of labour and through adaptation, has its own role to play. Thus some catch fishes, and these are passed on to those which function as mouths and stomachs feeding the rest of the colony. Other specialists are the float with its sail, and those concerned with reproduction. In the photograph a fish has been drawn up to the feeding polyps.

The family of jellyfishes includes, among its many different species, dwarfs and also giants attaining six feet in diameter and weighing over a hundred pounds. Most are motile, but there are sedentary ones, which hang from algae like chandeliers. From their fixed positions they use their tentacles to catch the innumerable planktonic animals which are the basic diet of very many small marine carnivores.

From Norway to the Gulf of Gascoine, the 15-spined Stickleback *(Spinachia spinachia)* nests among seaweed.

Nest-building fishes must be considered highly advanced, for in water more than anywhere else, care of the young is a good indication of evolutionary progress. Consider the difference between the ordinary Sardine which must lay thousands of eggs in order to perpetuate the species, and the female 15-spined Stickleback which need only deposit a hundred eggs in a nest built by the male. The spawn which the Sardine lavishes on the open sea is a ready-made pasture for hungry mouths, whereas the "brooding" which the Stickleback supervises with such care has every chance of success. But in the final analysis, each species has roughly the same number of surviving descendants. In April the male Stickleback prepares for the coming events. A clump of seaweed becomes the centre of his domain, from which he will chase all intruders. The possession of a territory induces a high state of excitement in the male, and he soon adopts a striking colouration and starts bending the branches of seaweed. He glues them together with a sticky secretion, intertwining them into walls and packing the holes with pieces of seaweed. The two freshwater species have 3 or 10 spines, whereas the marine Sticklebacks have 15 needles.

Once the nest is complete, the Stickleback finds a female and escorts her back to do her duty in egg laying.

The little female is bulging with ripe eggs, but as soon as she has deposited her precious burden she loses all interest. Thereafter it is the male that does all the work: he pours his milt over the eggs and during a fifteen day vigil fans his treasure energetically. His task continues even when the little ones hatch out, for they are denied access to the wide world for a little longer: the fresh-water Stickleback even takes adventurous youngsters into its mouth in order forcibly to return them to their nursery. Some other fishes also build nests. The male and female Wrasses *(Labridae)* build theirs together, and both guard the offspring in a shallow depression of sea-weed and broken shells. The Blennies and the Miller's Thumb too guard their spawn in crevices in rocks. Both parents of the Gunnel or Butterfish *(Pholis)* wrap them-selves round the mass of eggs which is rolled into a ball. Some of the most amusing to watch are the Gobies which will squabble violently over an empty shell. The winner uses it for a shelter, making a hollow underneath. Females are enticed to enter and attach their eggs to the roof of this cavern, and then the male camouflages the hiding place with sand and guards it for 9 days.

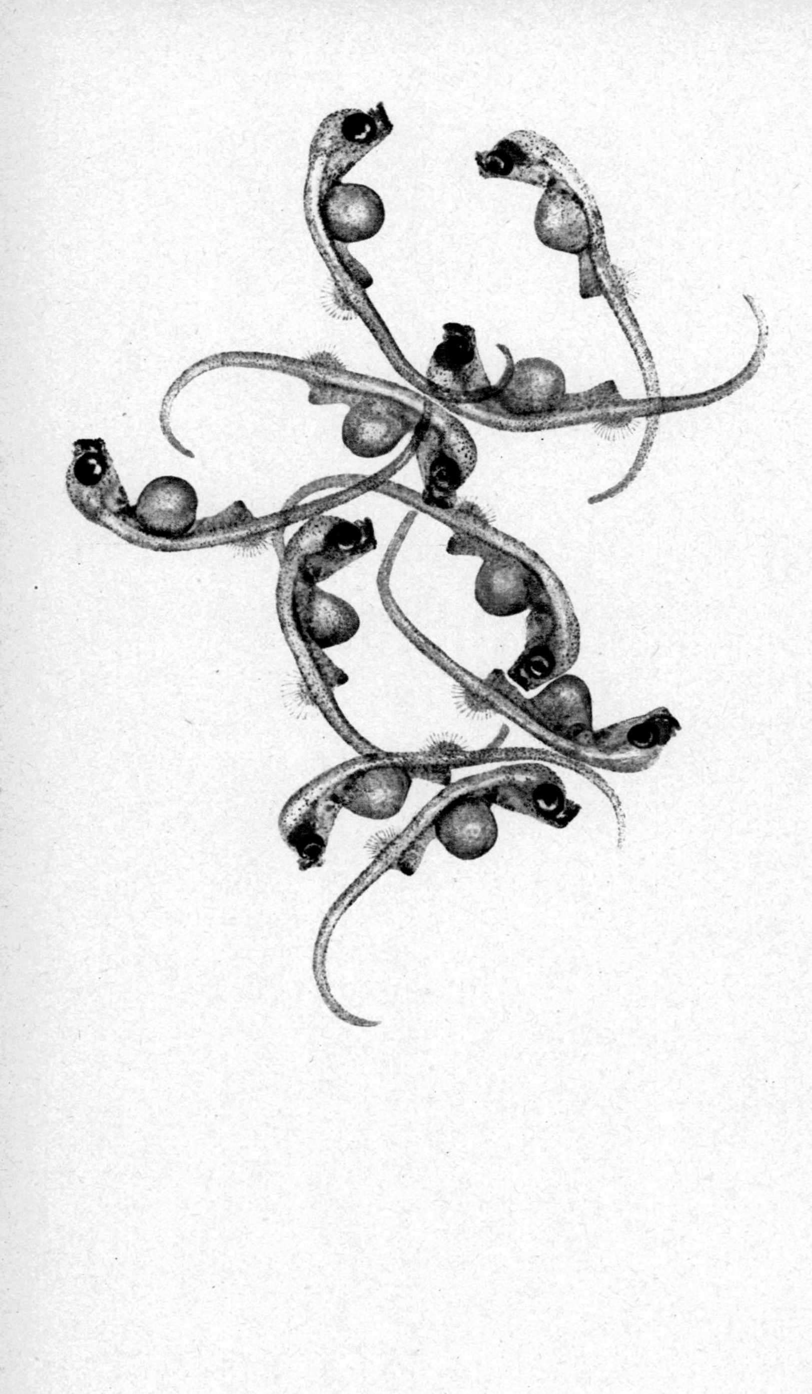

The Seahorse is a most original fish, both in appearance and breeding habits. Its head is at right angles to its body, and it swims by means of vibrations of the dorsal fin. In autumn the males and females congregate amongst the vegetation and perform a kind of nuptial dance. They rise and fall in the water, gracefully inclining and straightening, all the time getting closer to each other until, cheek to cheek and tails entwined, the pair embrace. Suddenly the female releases her eggs into the brood pouch of the male — the reverse of the normal procedure. The male seahorse carries them in his "womb", and further females will come to him with their eggs. Protected in this marsupial-like pouch, which is richly supplied with vascular tissue, the embryos will develop without even coming into contact with sea water. As the young grow, the belly of the male becomes distended, but when his brood of perhaps a hundred becomes too restless, the male must begin his "confinement". Taking hold of something for support, he twists and contorts his body until the pouch opens and, a few at a time, his progeny are able to escape.

➤ The Seahorse *(Hippocampus guttulatus)* attains 5 ins in length. It belongs to a family, the Syngnathidae, which incubate their eggs in pouches. It feeds on plankton, and strangely enough has practically no enemies.

⅄ While in the incubation pouch of the male, the young fry have a large yolk sac. On leaving the pouch they are $^1/_8$ inch long.

Some underwater regions simply explode with a riot of colour, just like the most beautiful terrestrial gardens, and so it was that the first divers gave the anemones the names of flowers. For a long time they were thought to be forms intermediate between animals and plant, but in 1723 Dr. Peyssonel of Marseilles discovered that these strange creatures were really carnivorous animals. They feel the approach of their victim with their crown of tentacles. They then paralyse it by means of their sting-cells, and by contractions of their powerful muscles they manage to suck down into their "hollow gut" or coelenteron quite fair-sized prey. At low tide they contract onto their attachment, leaving nothing more on the rock than an unrecognisable purple mass. They follow the tidal rhythm, and it is a curious fact that they maintain this rhythm even when kept in an aquarium, as if, in spite of the absence of a brain, they are able to retain the memory or distant perception of the tidal regime.

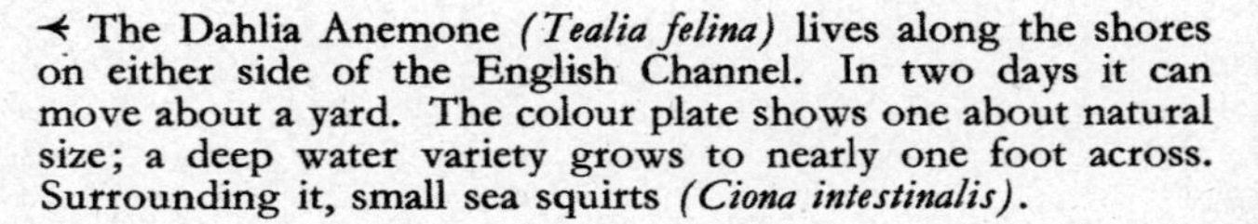

◄ The Dahlia Anemone *(Tealia felina)* lives along the shores on either side of the English Channel. In two days it can move about a yard. The colour plate shows one about natural size; a deep water variety grows to nearly one foot across. Surrounding it, small sea squirts *(Ciona intestinalis)*.

► The Beadlet Anemone *(Actinia equina)* digesting a small Wrasse. This anemone, found along the Atlantic and Mediterranean coasts, often survives for 15 years in an aquarium. It incubates its eggs in its intestine, from which emerge little anemones just like the parent.

An empty whelk shell does not remain "to let" for long. Large or small, there is always a Hermit Crab the right size applying for lodgings, since at each moult it outgrows its old shell. When the Hermit decides to move house, it first of all explores carefully the inside of its new abode. If no other creature is present, it slips in as soon as it has withdrawn its soft, pink body, so tempting for any passing predator. But hardly has it taken shelter than it realises that it has left something behind — the little anemone! Therefore, with its pincers it detaches its friend from the roof of the old house and gently instals it on the new one. The anemone acquiesces, because they have "known each other" since the young hermit crab took possession of its first shell and the anemone moved in too. Thus they live in a productive form of symbiosis. The Hermit carries the sedentary animal about with it, enabling it to find various kinds of prey. The anemone, in return, uses its stingcells to protect its host, and this defense is by no means unnecessary, for if the shell was truly as impregnable as it appears, then the original builder would still be there.

A Hermit Crab *(Eupagurus bernardus)* in a whelk shell. Sharing the shell are an anemone *(Calliactis parasitica)*, and a Nereid Worm, visible at left.

Fortunate are those who do not require a builder or contractor to build their houses! The Cerianthid, using mucus exuded from its skin, cements sand particles into a protective covering. Its long stinging tentacles show its affinities with sea anemones and other carnivorous anthazoans. The little "flowers" nearby, on the other hand, live in flexible though stiffish tubes of cemented mud particles, and they display merely the inoffensive gills of tube-living annelid worms. Cilia on these little fans waft towards them the plankton on which they feed, but even a shadow or slight movement nearby is enough to send them back instantly into their long tubes, and in a flash the "garden" is transformed into a grove of bare sticks. Then, after a few minutes, the shimmering plumes reappear. If there were a prize for speedy reflexes, it would surely go to these tiny marine palms, for even the fastest fish is no match for them. And yet a fossil from the Lias stratum in Wurtemberg shows a 3 ft. high ancestor of these worms that was caught with gills outstretched when a marine catastrophe struck with unbelievable swiftness 180 million years ago when Central Europe lay under the sea.

A Cerianthid *(Cerianthus membranaceus)* whose cylindrical stem can grow to over 2 ft. and the crown to 10 ins in diameter, among some Fan-Worms *(Spirographis)*.

A tussel between an adult Spider Crab *(Maja squinado)* and a young Spiny Lobster *(Palinurus vulgaris)*.

The sea abounds with crustaceans and their larvae, but most of them measure only an inch or so and can serve as a meal for only the smallest gourmets. Those creatures with larger appetites, the octopus or man himself, prefer crayfish or lobsters to these small fry. Along the rocky coasts they are caught in water of 25 fathoms. During the breeding season, however, it is necessary to give them some respite, for their development is long and complicated. For some months the female carries her eggs suspended under her tail: the larvae which hatch out must metamorphose four times before finally assuming the true crayfish form. But only after many months and numerous moults do they attain a length of 10 inches and are themselves able to reproduce. — In this startling photograph the young Spiny Lobster is dwarfed by the larger Spider Crab, which is not as fierce as it looks, although it can grow a carapace measuring 12 ins. Its long pincers make excellent weapons and also delicate utensils by means of which the crab gathers small sponges, seaweed or colonial polyps which it places on top of its spiny carapace where they continue to grow. So disguised, the Spider Crab can wander through the underwater jungle without arousing too much suspicion.

A 3½ ins Shore Crab *(Carcinus maenas)* in the act of opening an edible mussel *(Mytilus edulis)*.

The macrourid crustaceans — lobsters, crayfish and spiny lobsters — are climbers as well as swimmers. They can propel themselves along, though only backwards, by beating their tail fin. The anamourid crustaceans, such as crabs, however, can only progress by using their feet. Crabs vary greatly in behaviour: some are lazy, like the Swimming Crab, which goes its solemn way, topped by a spongy canopy; others are petulant, like the Green Crab shown here, whose insolence and sheer cunning are unrivalled. The latter devour anything which falls near their pincers, but at the least sign of danger they will disappear in a flash into a crevice. Their favourite dish is mussels, although it takes them some time to force their shells apart. Often the mussels themselves harbour two parasitic Pea Crabs *(Pinotheres)* which lose their home through the more powerful relative's attack. While *Pinotheres* is one of the smallest of the crabs, *Cancer pagurus* can weigh 12 lbs. or more. However, crabs do not just grow: at each moult they must vacate the old carapace with all their paraphernalia of pincers, claws, mandibles, antennae and stalked eyes. Until the new carapace hardens, even the most aggressive crabs must hide and fast for a while.

A battle between hard and soft: in spite of its hard carapace, the Lobster *(Homarus vulgaris)* will lose.

The octopus relishes the delicate meat of crabs and shellfish, but requires a great deal to fill its large stomach. That is why it prefers to make a meal of lobsters, even though their capture is sometimes dangerous, for the lobster is fearless and well aware of the strength of its pincers, having already tested them on the limbs of rivals or in opening mussel shells. Fiercely counter-attacking the octopus, it pinches the adversary's tentacles, but realises too late that these leathery muscles are slipping out of its grip. Holding its ground, the octopus merely shows its rage or pain by a change in colour. Struggling for survival the lobster then tries to pierce the balloon-like body of the mollusc, pinches it but is deflected, as the angry octopus releases the contents of its ink sac and the two combatants are enveloped in a black cloud. The bodies become lost in darkness, where the eight arms of the octopus writhe, trying to attach their suckers to the armoured carapace of the lobster: the embrace tightens and when the cloud slowly lifts, the fight has ended.

The Lobster's claw can now only snap at the water: the Octopus *(Octopus vulgaris)* has crushed its victim.

The octopus has slid on top of the lobster and holds it prisoner under its mantle. Below, in the middle of a vault of tentacles, is its mouth, with jaws resembling a parrot's beak. Now, the octopus breaks the carapace of the lobster at the joint behind the head and injects a paralysing venom into the wound. Then with a rasping tongue *(radula)* like a chitinous file it scrapes out the flesh from inside the carapace, all the while keeping a watchful eye on its surroundings. — With good luck or prudence, a lobster might live for 50 years, but, like many others, it may fall a victim to the well-baited lobster-pots which, along the coasts of Heligoland alone, catch some 60,000 lobsters a year. These crustaceans, so abundant in northern waters, are rare in the Mediterranean. There, the traps attract Spiny Lobsters *(Palinurus vulgaris)* . . . or octopuses raiding the pots. Although feeding on the delicate flesh of crustaceans, an old octopus makes a tough, leathery dish. The fishermen therefore prefer the young ones, which they catch at night with a trident and lantern.

Octopus vulgaris. The body of an adult measures about 6 ins and the arms 16 ins each. Giants 9 ft. in circumference are rare.

Compared with the Hydra of Lerne or the giant octopuses of Jules Verne, real octopuses are rather small. All the cephalopods are molluscs, related to the gasteropods and mussels. In the Octopus no trace of a shell remains, but the Cuttle-fish has a small internal shell under its skin, the "cuttle-fish" bone that is given to cage birds. Its relative in warmer seas, the *Nautilus*, has a pearly shell, while in the Mediterranean the female *Argonaut* (20 times larger than the male) has a slender receptacle where she deposits her eggs. The Octopus having abandoned outside protection, relies on the acuteness of its senses, its ability to adapt to its background, and its ink sac. It spends its days in a crevice or behind a barrier of rocks which it has built itself. It only travels at night, touching its legs on the sea bottom like stilts, or clinging to the rocks and hauling itself along. Its worst enemy is the ferocious Moray Eel, from which it escapes by emptying its ink sac and violently expelling water from its body to propel it forwards, its legs trailing behind. The Moray Eel often gives up, but takes advantage of a female Octopus whose maternal instinct keeps her in her lair, near her innumerable eggs.

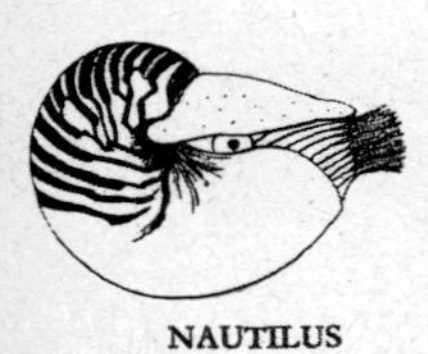

NAUTILUS

Cuttlefish *(Sepia officinalis)*, cephalopods about 1 ft. in length, have ten arms, two of which are rectractile.

Legends tell of giant octopuses, but these monsters were in fact very large squids, which are known to reach the colossal length of 60 ft. The Cuttlefish deserves the title "sea chameleon". Its eyes, which are extremely complex, transmit all changes in the colour of the surroundings to its nervous system, and without a second's delay, thousands of chromatophores on the body take on the same shade as that part of the sea bed on which the cuttlefish rests. During the summer mating season, however, when a male sights a female it assumes a pattern of brilliantly contrasting stripes *(see photograph)*, and the female replies in the same manner. At night their stripes are enhanced by the iridescence of their phosphorescent skin, and their love play looks like a ballet as tentacles interlace and bodies turn. Soon the females deposit some hundreds of eggs on seaweed, corals, or floating branches. These are often called "sea raisins", since the eggs are distributed on the branch like a bunch of dried grapes *(see page 84)*. After egg-laying, the female does not bother further with her progeny, for as soon as they leave the egg, the young cuttlefish can, just like the adults, change colour and spit out ink.

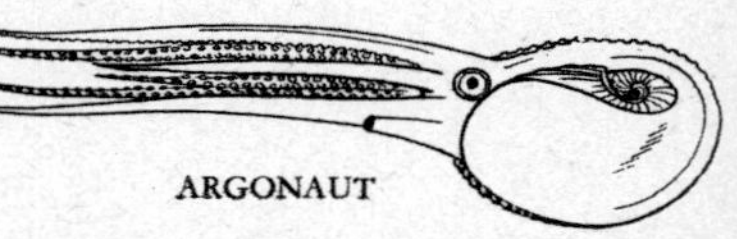

ARGONAUT

The Loggerhead Turtle *(Caretta caretta)* attains 3–4 ft. in length, and may weigh nearly 200 lbs.

Real living fossils! Two hundred million years ago their ancestors, closely resembling modern turtles, swam and crawled beside ichthyosaurs and brontosaurs. While many strange species became extinct, the turtles have survived and spread into all tropical and temperate zones. They are mostly carnivorous, although a few are herbivorous. The large Loggerhead Turtle still haunts the Mediterranean, since its flesh and carapace are less sought after than those of its relatives, *Chelonia* and *Eretmochelys*. With flippers like oars, turtles are excellent swimmers, coming to the surface to fill their lungs with air or to bask in the sun. They feed on shrimps and fish, and get rid of excess salt through glands in their eyes and nostrils. Only at egg-laying time do the females leave the sea. On the beach they laboriously trace a path in the sand. In hollows two feet deep they deposit 60–150 eggs; then they carefully cover them up and the heat of the sun does the rest. During this incubation period, the eggs draw from the damp sand the moisture which they need. Then, after a month or two, if the eggs have not been stolen by animals or men, the minute turtles hatch out, dig their way to the surface at dusk, and instinctively make for the sea.

The freshwater European Pond Tortoise *(Emys orbicularis)*, found in the south and east of Europe. Carapace 1 ft.

Can one believe the stories of the fabulous age reached by the giant tortoises of the tropics? Some carried, engraved on their carapaces, dates (e.g. 1766) furnishing evidence of their age. But even the little freshwater tortoise could certainly live for a hundred years. Such high life expectancy is due to its good armour and careful ways, but also to its small expenditure of energy. The metabolism of animals with variable body temperature *(poikilotherms)* is much slower than that of warm-blooded animals *(homoiokilotherms)* which must above all maintain their optimal temperature. Consequently, from a hundred years one must deduct a hundred full winters which the tortoise passes buried in the mud, almost without breathing, barely alive. It awakes in spring, to hunt small fish, larvae or frogs. Breeding is preceded by wild fights between rivals. When at length the male conquers the female, they cling to each other for several days. In June, the female climbs the river bank to bury 3–16 eggs... but why do aquatic tortoises deposit their eggs on dry land? In the course of ages, their ancestors left the water, to live on the land. During an ensuing period of flooding, certain species were forced back to an underwater life, but still retained some of their terrestrial habits.

With the first rays of the March sun, the newt abandons its lethargy and slowly leaves its hole to return to the liquid element. Scarcely has it entered the water than its awkwardness vanishes and it gaily avails itself of the myriads of tiny animals which teem around it. In the breeding season, the Alpine Newts assume bright colours, the smaller male displaying blue flanks in contrast to its orange belly. Once conquered, the female begins to build "cradles" of algae in which to hide the numerous eggs. Egg-laying occurs at night: with her small "hands" she bends aquatic plants into an arch, gently so that they remain smooth and supple, and in each little basket she lays an egg. In this way her offspring is better protected than frogspawn floating at the surface in easy reach of greedy mouths. After two or three weeks the larvae emerge, but it is still another three months before they reach the metamorphosis which makes them adult newts, complete with lungs. Following this, successive moults are accompanied by increase in length and, so that nothing is lost, the old skin is immediately eaten. If an enemy bites off one of the newt's feet, that much body tissue must be made good. For in the newt, everything can regenerate, even the jaws and the eyes!

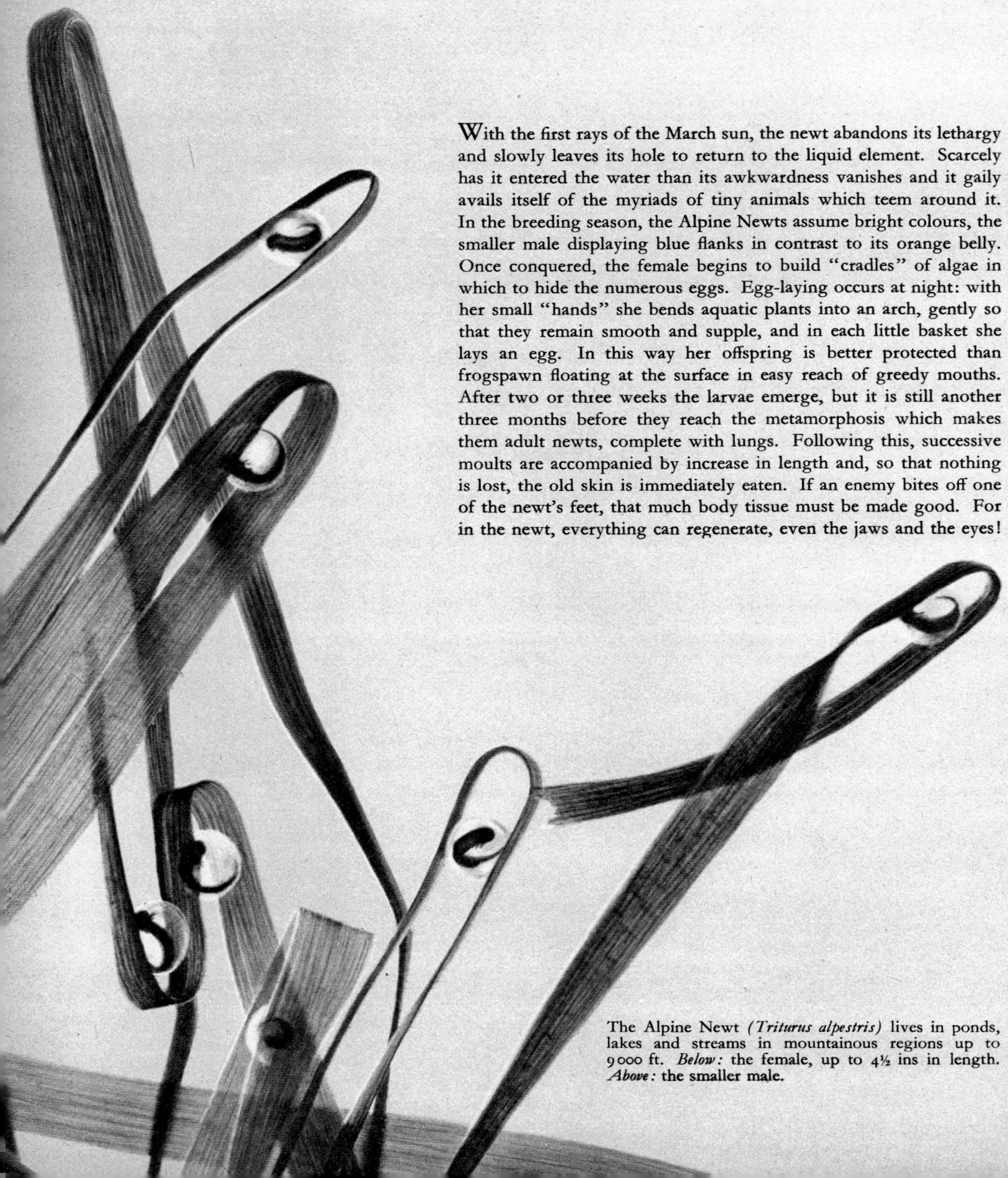

The Alpine Newt *(Triturus alpestris)* lives in ponds, lakes and streams in mountainous regions up to 9000 ft. *Below:* the female, up to 4½ ins in length. *Above:* the smaller male.

Except for the Viper, European Snakes are almost harmless. Out of 400 species of poisonous snakes only 9 inhabit Europe. The Dark Green Snake or European Whipsnake belongs to the non-venomous family of the *Grass Snakes*. Only *Vipers* or *Adders* produce venom with power to disintegrate the blood cells, the poison being squirted through the hollow fangs of the upper jaw. They use this weapon to paralyse their prey quickly or to protect themselves against attacks. But these very snakes, Vipers and Aspic Vipers, use flight as the best form of defense. There is little chance of meeting them as they only emerge during the daytime in spring, when they warm themselves in the sunshine after their long hibernation. They are nearly blind in sunlight for their eyes are better adapted to a crepuscular existence. In fact, snakes spend more time in their lair than outside! The horny layer of outer skin does not grow and every two months becomes too tight; the snake becomes feeble and inert and its eyes cloud over, for even the cornea is shed. Utterly incapable of hunting, it hides and fasts for days, and when the skin has loosened round the mouth the snake rubs itself against something until its head is freed and throws off its transparent skin like a stocking. Only then does it come out to hunt again. A snake's skin is very elastic and covered with small scales. The species can be identified by the arrangement of scales on top of the head. The Greeks believed that anyone finding a sloughed snake-skin would have a long life.

← A dramatic scene photographed in Northern Italy: A Dark Green Snake *(Coluber viridiflavus)*, 6½ ft. long, encircling a young Green Lizard. An adult lizard is up to 21 ins long and can attack young snakes in its turn.

After shedding its skin many times the Southern European Aspic Viper *(Vipera aspis)* reaches a length of 25 ins. North of the Alps it is only to be found in the Black Forest and in the Jura in Switzerland. →

The photographer, waiting in his hide for the Woodchat Shrike *(Lanius senator)* to return with food for the young birds...

Many dangers surround every bird's nest. *Danger* is not the exception in nature, but *survival:* only 10 out of 100 fledglings live to see the next spring. Innumerable predators make their way through the tree tops... martens, snakes, and above all crows, magpies and jays. The very Shrike whose brood is being attacked by the snake today, probably plundered a song-bird's nest yesterday. The song-birds will rear a second brood and the female Shrike will also lay again. Life goes on.

The Aesculapian Snake is very well adapted to climbing trees. With its breastplate firmly pressed into every notch in the bark, it can climb the tallest tree and often sleeps high up in the fork of a branch or in a hollow tree. It lives in the light deciduous woods of Southern Europe as far north as the Balkans and is occasionally found in the Black Forest and the Taunus, where in Roman times it was set up as the symbol of Aesculapius, the god of medicine. It does not occur in Britain.

...witnessed the seizing of a nestling by a 6 feet Aesculapian Snake *(Elaphe longissima)*.

When hunting the Snake does not rely merely on sight. Its whole body registers the slightest movement and its tongue, which is the organ of both taste and smell, senses the presence of prey, even young Shrikes lying still in their nest; and the noisy protest of the returning adult birds will not deter the snake once it has started the raid. — Later in the year, when the young birds are fully fledged and the nests abandoned, the Aesculapian Snake has to seek other food and turns to catching mice and lizards. It hibernates until late in May; then at the end of June lays 5 to 8 eggs which it hides in humid soil. In early autumn the young snakes, 8 to 8½ ins long, hatch and have to fend for themselves. These, in their turn, fall victim to countless enemies, for they are helpless against the huge Mediterranean Eyed Lizards, adult Adders or Vipers, Polecats, Hedgehogs and Short-toed Eagles. Thus the life cycle is closed; in nature every animal is at once hunter and hunted: every living thing lives on other living things to keep itself alive.

Tadpoles are the food of numerous carnivores, amongst which are young Grass Snakes *(Natrix natrix)*.

At the end of autumn, little six inch Grass Snakes leave their parchment-like eggs. Far from fearing water, they make their way without delay to the nearest lake and there hunt tadpoles and other batrachian larvae. However, hibernation slows down their growth and at the end of two years they scarcely measure eighteen inches. But if they live for ten or twelve years they may attain six feet. They swim in water with ease, head high like the prow of a ship, but plunging down to catch their prey. On land, they show themselves agile enough to cope with frogs. A large Grass Snake can easily swallow a dozen one after the other, without chewing, and line them up in its distended oesophagus. Jaws and larynx are so supple and extensible that the snake never suffocates during this laborious process of ingestion. Having fed, the snake goes to earth in a hole, and no longer shows any craving after flesh, but conscientiously digests. Its gastric juices can dissolve even bone: from a mouse it has swallowed, it rejects only fur and claws. Should it be attacked itself, it feigns death, and from its stink glands exudes secretions in order to disgust its attacker. Then, when it judges the danger past, it takes off to the water for a purifying bath.

A well-stocked larder: in a mass of frogs' spawn these Alpine Newts *(Triturus alpestris)* take the ripest eggs.

Frogs do their best to propagate their species, but nature imposes certain restrictions. In May, as soon as the female has laid the eggs, the rapid transformation begins, from a unicellular to a highly differentiated organism. In six days they pass through what took some millions of years of evolution. During this brief time, thousands of eggs fall prey to the gluttony of newts, and even of parent frogs. As tadpoles one or two weeks old, not yet with legs, they still continue to be eaten in their thousands, even before they have been able to feed on algae, their first food. After three months the gills cease to be visible from the outside. The hind legs push through, and the front legs develop in little sacs in the skin. After another month the tail atrophies and lungs develop. At the fourth month, it is a tiny but perfectly formed frog which leaps out onto the bank; henceforth it will breathe from the air and not under water. From 10,000 eggs laid by the female, only a few individuals will reach adulthood, and, four years later, themselves be able to reproduce: these are the most agile or the most prudent, the fortunate ones which have managed to escape their enemies, Pike, Grass Snakes, Polecats, Water Rats, Storks or Herons.

A common drama in still water in Central Europe: Water Boatmen *(Dytiscus marginalis)* attack a frog. Magnified.

Many terrestrial animals have become adapted to life in water. The Water Beetle or Boatman not only skims the surface in its slow flight, or crawls on the land: it also swims and dives. Its hind legs bristle with fine hairs and act as oars. Taking a store of air in its tracheal tubes and under its elytra (or wing cases), it follows its prey under water: tadpoles, centipedes, small fishes and young frogs. Its sharp jaws never relinquish its prey, and often a second beetle comes to its aid while the first quickly makes for the surface to breathe. Their breeding habits are quite violent: the male seizes the female round the "neck" with his anterior adhesive pads, and for some days whirls around with her. Following this the female uses her ovipositor to bore into the leaves of aquatic plants and deposits about a hundred eggs. The carnivorous larvae climb onto dry land at the end of three months in order to metamorphose. When the perfect insect emerges, it once more plunges into the water.

Running water, however, is the hunting ground of the Water Shrew *(Neomys fodiens)* – here magnified 1½ times.

A mammal, whose blood must be kept warm, requires special adaptations if it is to be able to spend its life in water. The Water Shrew possesses thick oily fur and a special fan of hairs between its claws and at its tail which help in swimming and steering. Like a tiny otter it swoops into the middle of a shoal of small fish or tadpoles, or running briskly along the stream-bed, it snaps at larvae, snails or leeches. After each catch, it rises breathlessly to the surface for air. If the occasion arises, it will not hesitate to attack fish 60 times bigger than itself. It will even leap on to the back of a carp and attempt to eat the less protected parts of the fish. After its meal, it makes for its lair through an underwater entrance and then leaves on the landward side to dry itself in the sun. But should a frog come near, it is off again, for owing to its extraordinarily high rate of metabolism it has an insatiable hunger and never hibernates.

Eggs and larvae of the Italian Spectacled Salamander *(Salamandrina terdigitata)* x 5. The larvae spend the winter in water.

Salamander eggs are rarely found in water. In fact the majority of salamanders occuring in northern latitudes have left the water, their ancestral home. But all of them need humidity and can tolerate terrestrial life only because their skin, richly provided with glands, secretes sufficient mucus to prevent them from drying up. The Black Salamanders of the Alps carry their eggs for 18–36 months before depositing on the dry earth two tiny salamanders, perfectly formed and equipped with lungs. The female Yellow-spotted Salamander carries about thirty eggs for some months, but in the spring puts them in the water: almost at once well-formed larvae hatch out, which spend another three months in water, breathing with gills. Both these methods protect the eggs and larvae from the chilly northern waters. In contrast the lakes and streams of central Italy harbour the eggs of Spectacled Salamanders. After an underground hibernation, they mate on the ground and the female then abandons the eggs to the warmth and dangers of the water.

Mediterranean Eyed Lizards *(Lacerta lepida)* hatching, enlarged. The eggs need a temperature of 79–82° to develop.

In a few years these tiny lizards will resemble the dragons of fairy tales. Only Liguria, Provence and Southern Spain have the right soil humidity for Mediterranean Eyed Lizards to incubate, for the temperature must be neither too hot nor too cold. So these gorgeous lizards, the biggest in Europe, are only to be heard hissing here and there in cracks in the cliffs of the Western Mediterranean coasts, where they hunt their smaller relatives with great agility. They can also overcome young birds, mice and snakes, and when frightened will even spring at the throats of wandering cats or dogs. If these warriors were to attack their rivals with equal rage at mating time, they might well exhaust themselves or kill each other... but for a preventive behaviour mechanism hereditary in many well-armed species. Hence the battles between the giant lizards have dwindled into a mere ritual: the stronger one simply threatens the other and seizes it by the head symbolically, without biting it. With a ritual show of modesty the defeated lizard then takes its leave.

The largest European representative of the modern *Sauria* is Pallas's Glass Snake *(Ophisaurus apodus)*, not a true snake.

The red-brown or straw-yellow Glass Snake, which has existed since pre-historic times and is over a yard in length, still slithers through the undergrowth in the valleys of the Balkans. Although it resembles a snake in appearance it actually belongs to the lizard family. Like the Blind Worm, the Glass Snake prefers to hunt for snails, lizards and even mice in the twilight. As soon as it seizes a larger animal, it turns round and round and round until its victim is giddy. But if it itself is attacked it cannot, unlike the other Sauria, leave its tail between the enemy's teeth. Instead of biting back, it squirts out the contents of its intestines if necessary. Then this apparently legless reptile glides away from the scene — only on close examination can one see the tiny stumps or vestigial hind legs, which no longer serve any purpose, but which every Glass Snake possesses on hatching out of the egg. The females leave their eggs to incubate in the rays of the sun, while the Slow Worm (or Blind Worm), which lives further north, incubates its 6 to 12 eggs in its own body, protected from changes in temperature.

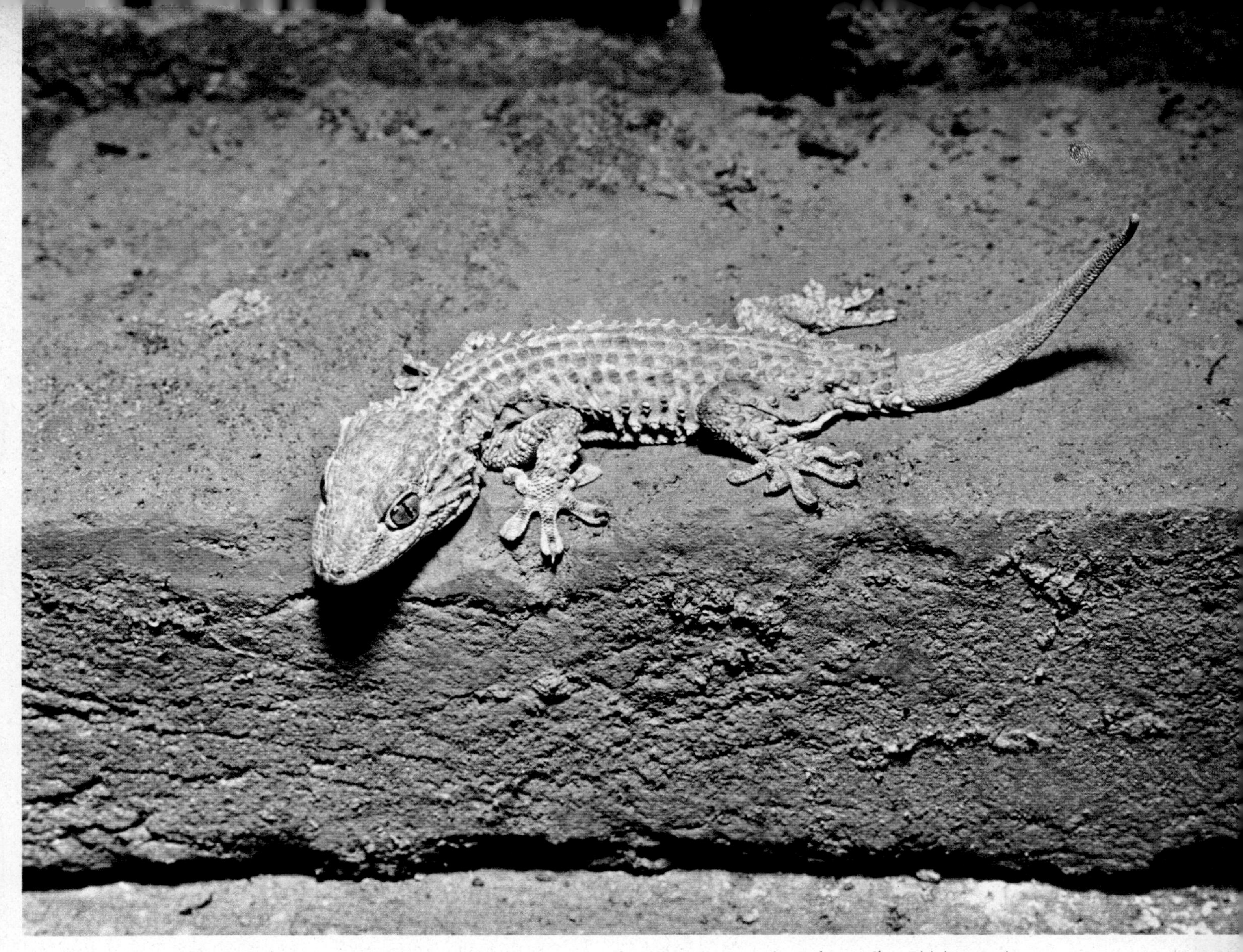

Wall Gecko *(Tarentola m. mauritanica)* with new growth of tail. Geckos are the only reptiles which squeak.

Round the Mediterranean and on its islands, Wall Geckos are to be found in the evening hunting insects on masonry and ceilings. They have broad, flat fingers and toes with claws on the third and fourth toes, and adhesive pads on the underside which enable them to stick where there is the slightest unevenness. They creep along over vertical surfaces and ceilings, stalking their prey like cats. Then suddenly they will spring and catch a moth, but when frightened they escape at tremendous speed. Although they do emerge during the day, they are more active after sunset, when their huge night eyes with narrow pupils are especially useful. Their lower eyelid has formed a transparent film over the eye. Wall Geckos are often to be found in gardens and in houses and are welcomed by man for they also catch flies and midges. They vary in colour from grey to blackish brown and yellow, and squeak a great deal before pairing. The female lays two eggs in some sheltered hiding place and the young hatch after 4 months. They are independent from the moment of hatching and can shed their tails should an enemy seize them. Geckos are smaller than lizards, the Wall Gecko being 4–5 ins long.

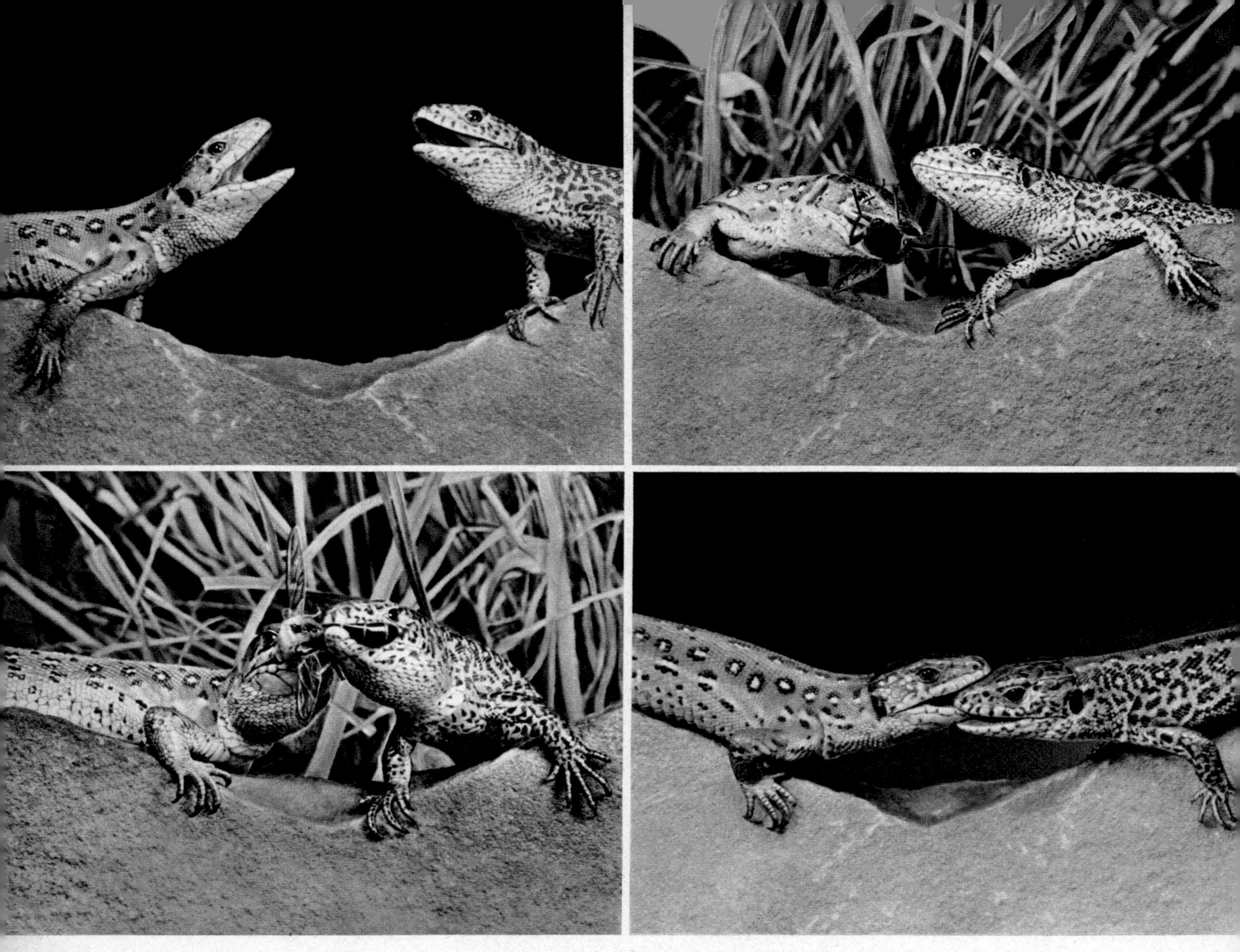

Sand Lizards *(Lacerta agilis)* before pairing. Left, the female and right, the male 6 to 7 ins long, with a tail more than half its length.

When the Sand Lizard comes out of hibernation in early April, its colouring is of a brownish purple. But with the beginning of the courtship season, the green colour of the male becomes decidely more pronounced and the underparts become yellow with black dots. From May to June the male is very active, for he has to secure a territory and to defend it against intruding rivals. Once he has persuaded a female to share his hunting grounds, he treats her nicely and lets her eat all she catches, and if *he* makes a good catch, he will even allow her to nibble at his prey. Then, in early July, the female looks for a patch of loose soil where she digs a hole for the eggs. She lays 5 to 14 white eggs with shells resembling parchment in texture and covers them over with leaves or sand to be incubated by the sun. The young hatch in late July or early August and are 1½ to 2½ ins long and grey-brown in colour. In October the Sand Lizard goes into hibernation, spending the winter underground. Its habitat is on heathland, sand dunes and fields and glades, from Western Europe to Asia; in Britain they are found only in some Southern counties and not in Scotland or Ireland.

In the autumn sunshine young Green Lizards *(Lacerta viridis)* are lying on the back of an adult lizard.

Before the winter hibernation, which lasts for three to four months, Green Lizards make as much of the autumn sun as possible. In this season, reptiles which would normally hunt each other, can be seen huddling together for warmth on stones and young ones may even creep out of the cold grass onto the back of an adult. On autumn mornings the lizards awake nearly frozen and are inactive until their body temperature reaches 85–90°. Small bodies warm up faster than big ones, and thus the young will have made off before the adult gets active and hungry. It is a narrow escape, for these large Green Lizards do not only eat insects, eggs and frogs, but also small lizards and even their own offspring. They are 12 to 20 ins long and their skin is covered with a type of scales known as tubercules. The skin is shed periodically and the lizard changes in colour after the moult. The female is green all over, but the male has a blue throat during the breeding season. Unlike the snakes, the lizards have mouths which cannot expand and they walk or run, dragging their bellies along on the ground.

Male and female Green Lizard *(Lacerta viridis)* prior to pairing, when tail biting takes place.

The display ceremonial of the Green Lizard is a highly complex one. First of all the male approaches the female and lifts his head to show his azure blue throat. The female looks at him, shakes her head and approaches him hurriedly. She is hesitant, yet at the same time precipitate. If she wished to reject his advances she would stamp heavily and twitch her long slender tail. If, however, she accepts, she allows him to take her tail in his mouth. After the male has been following the female for some time, she will stop warding him off and pairing will take place. – If a lizard wishes to escape from another animal it can shed its tail — for there are points at intervals along the vertebrae of the tail, in the connective tissue and even in the blood vessels, which can be broken without any difficulty if there is danger. The actual breaking is not effected by pulling, but is due to a muscular contraction set in motion by a shock. Afterwards, the lizards can be seen running about with the stumps or a new-growing tail. Green Lizards sometimes escape from captivity to live wild in the British Isles, but the species is not native, though it, as well as the Wall Lizard, are found in the Channel Islands.

One of the Skinks *(Chalcides ch. chalcides)*. Its slate-grey to olive-green skin has a metallic gleam. Up to 17½ ins long.

The Skinks are an unusual family of lizards, resembling the Slow-Worm but with the shrunken limbs of a lizard. They are frequently found in sandy places and deserts, where they catch scorpions, but this particular Skink inhabits damp meadows, chasing its prey at great speed by winding through the grass. The many different species of Skink show as many different stages in the development of reptiles, some having very short limbs like the one in the photograph. In the Round-bodied Skink *Chalcides ocellatus*, the front pair of legs is very small and can be withdrawn into a groove in the body. Some like the Greek *Ophiomorus punctatissimus* have no limbs at all, and others like the *Ablepharus kitaibelli* living in Roumania and Hungary, have long limbs. The female *Chalcides ch. chalcides* usually produces about 15 young. This Skink eats insects and snails and is found in Spain, Southern France, Italy, Sicily and Elba. The limbs are about $^1/_{10}$ to $^3/_{10}$ inch long and the three fingers and toes have tiny claws, but these are not visible in the photograph. The reasons for this mutation of the limbs of many Skinks are unknown; we can but wonder why the species which have adopted snake-like motion still preserve these rudimentary limbs which are no longer of any use.

Female Tarantula *(Lycosa tarentula)* with young, enlarged 3 times. Its dreaded bite is no more dangerous than a bee sting.

Centuries ago it was believed in Taranto that anyone who was bitten by the Tarantula was already as good as dead. Wildly gesticulating, he would dance to the strains of the Tarantella, to prevent himself falling into "everlasting melancholy". Thus the legend of the murderous Tarantula rapidly became widespread throughout Europe. There are over 20,000 species of spider scattered throughout the world and on each average square yard of open country at least 50 spiders of all sizes lie in wait for prey. Their usefulness to man is greater than that of the birds, for they are all tireless hunters of insects, either lurking in their wheel-shaped webs or holes in the ground or creeping up and pouncing upon their victims. The Tarantula is one of the wolf spiders *(Lycosae)*, which do not set traps, simply lining their earth funnels 6 to 8 ins deep with their spiders webs. After pairing the females kill and devour the male which is their last sustenance before they devote themselves to their offspring. Having laid their eggs, they spin a cocoon which is fastened to their abdomen, thus enabling them to carry it everywhere with them. When the young hatch, they climb straight onto their mother's back, where they remain for some days without moving.

Young Scorpions *(Euscorpius italicus)* hatch immediately after the eggs are laid and climb onto the mother. (x 3½.)

It has been shown by the discovery of fossils that there were already scorpions crawling about the world 400 million years ago. There are 600 species in existence today, and it may be true that some exceptionally large ones are dangerous to man, but the Southern European Scorpions do not justify the fears of people holidaying in the country. They lie in wait for other insects and it is very interesting to see a scorpion preying on a centipede. It first gets into the attacking position and suddenly seizes the victim with its pincer-like front legs. Holding the twitching prey up towards the tail, also raised, it injects its poison. During the display ceremonial the pair stand with their abdomens turned upwards towards each other, entwine their formidable tails, release each other, and then catch hold of each other again by their pincers and dance until pairing takes place in a dark cranny. It frequently happens that the female devours the male immediately afterwards. Out of the eggs there hatch 20 to 30 young scorpions, which cling to their mother right up to the first moult. *Euscorpius italicus* is also common in the South of France.

Southern European Dry Wood Termites *(Calotermes flavicollis)*: soldiers, working larvae of different ages and a winged sexual insect.

Termites — also known as White Ants — feed on wood, construct marvellous nests and are much studied because of the societies in which they live. They avoid the light and do a great deal of damage to houses in Southern Europe. Where a royal pair of Dry Wood Termites founds a colony, however, not too much damage would be done, for their subjects consist of only a few hundred individuals, which are satisfied with decaying grapevines, tree trunks and posts. When whole palaces, museums and libraries are destroyed at the foundations and valuable works of art crumble at the touch, *Reticulitermes lucifugus*, the European Subterranean Termites, are at work. Their state contains hundreds of thousands of insects and they make so many tunnels and passages that it would be impossible in many cases to get rid of them without demolishing whole buildings, which would still not ensure that they did not reproduce. Even without a royal pair the colony can still survive and grow, for this species of termite can produce branch colonies with sexual insects, some females of which rapidly develop into substitute queens and together soon become as fertile as a true queen (8,000 eggs per day). 60% of all the larvae develop into workers, the insects which actually destroy wood: they digest the cellulose in wood with the help of their intestinal parasites and feed the rest of the colony with regurgitated digested food.

Common Wood Ants *(Formica nigricans)* surrounding Colorado Beetle 10:1. This pest arrived in Europe with American potato cargoes after World-War I.

An entomologist's explanation of the difference between termites and ants would fill a whole volume: but for the layman the main difference can be seen in the general harmfulness of termites and usefulness of ants. Out of 7,000 species of ant only very few feed on plants or wood; most of them prey on other insects and help to rid forest and cultivated land of pests. Ants also keep herds of aphids, or greenfly, which they milk, thus in the process protecting some plants from other insects' attacks. Some ants also collect seeds, often only nibbling the outer skin and leaving the hard kernel in the earth, thus spreading the seeds. There are ant colonies which live almost entirely on the secretions of other insects like the larvae of the *Lycenidae* (Blue Butterflies), Crickets or Staphylinia Beetles, and some which are looked after and fed by enslaved members of other species of ants. Even these ants are useful to farmers and gardeners, because they loosen the earth and aerate the soil. But of course all these activities are only unintentional by-products of the main work of each single ant which is directed towards serving the rest of the population and ensuring the survival of the colony. But the state could not survive if the territorial claims of every ant colony were not respected. This is possible because each ants' nest has its own distinct odour.

Great Green Grasshopper with a freshly-caught Grasshopper
(Tettigonia viridissima)

Some species of grasshopper, such as the Great Green Grasshopper or Bush Cricket, are occasionally carnivorous. Usually it feeds on plants but sometimes it devours smaller members of its own and related species. If an enemy, such as a lizard, approaches, the grasshopper immediately springs into the air — to fly a good 50 yards away. Grasshoppers produce their "music" by rubbing their hind legs over ridges on the abdomen. The Great Green Grasshopper has wings which are much longer than its body and the female has a long ovipositor. The antennae are long enough to reach the tip of this. It lives in trees or bushes.

Larva belonging to a species of Mediterranean Mantis
(Empusa pennata)

The particular mantis in this picture is climbing on foliage with its abdomen curled upwards. Unlike the grasshopper, the mantis with its very long supple legs does not produce any musical sounds. The praying attitude of all species of mantis is due to the unusual modification of the first pair of limbs, which have greatly elongated coxae, or proximal joints. On the curved underside of the femur of each leg is a channel with moveable spines on each edge. The tibia can be closed into this channel and the edges are well adapted to holding and cutting in order to maim the prey. The 3 chief genera of Mantidae are Mantis, Empusa, Eremophila.

Southern European Stick Insect, shown natural size
(Bacillus rossii)

An Italian Common Field Grasshopper, one of the Acrididae
(Acrida bicolor)

Compared with grasshoppers the Stick Insects might be called sluggish. They are so well camouflaged that, when they cling to a twig inert and silent, they appear to passing birds to be twigs themselves. They are diurnal insects and feed on juicy leaves. In the tropics there exist gigantic relatives of the European Stick Insect, up to 1 ft long. The most interesting of these is the Stick Insect widely used in laboratories, *Carausius morosus* : the wingless females of this species are capable of reproducing themselves without males and may be observed in insectaria. This parthenogenesis may continue for many generations, with females hatching from unfertilized eggs, even up to the 20th generation.

The Common Field Grasshopper is a vegetarian. It is a species of Short-horned Grasshopper, or *Acrididae*, all members of which have short antennae. Being few in number they are not as a rule dangerous to crops, but one related species, the Locust, is notorious for its devastation of vast stretches of land in different areas of the world. It still appears sometimes in swarms in Southern Europe and nowadays a continuous watch is kept on suspicious areas: in 1946 some 12,000 lbs of eggs were collected in Lombardy in 5 days and in other Italian provinces great hordes of locusts were destroyed by drastic means.

The Praying Mantis sits in wait for prey without moving, its arms raised as if in prayer, camouflaged among the grass. At the approach of an insect its grasping arms shoot out, and the insect is paralysed at once by the points on the ends of them. It is then greedily devoured, and its captor sits in wait once again. With the two large faceted eyes and three ocelli (or auxiliary eyes) on its brow the Praying Mantis can follow every movement in its surroundings. If a bird should sight it, it lifts up its transparent hind wings to make itself look dangerously large. The smaller male can fly — but for the heavier females the wings can only be used to frighten enemies. Yet in spite of their advantage very few males ever escape from the crushing arms of the female after pairing. After devouring her mate the female looks for a place which is sheltered from the rain, where she fixes her horny egg packets. The insects hatch during the following spring, and closely resemble the adult, except that they are wingless. From the moment of hatching they start hunting greenfly, and after several moults they are capable of overpowering larger prey.

The South European Praying Mantis *(Mantis religiosa)*, 2 to 3 ins long: ➤ Threatening pose. ➤ Moulting, pairing, catching prey, egg-laying.

Most people have at one time or another tried to pick a flower and in doing so grasped some "cuckoo spit" and wiped their fingers on a tuft of grass. But few of us know where this froth comes from. In each of the spumes of froth there is a larva of the Froghopper. It perches on the stem of the plant and sucks the sap. Its liquid excrements run down its back, diluting a wax which the larva secretes between the seventh and eighth segment of its body. Air which has been used for breathing is pumped into this soapy liquid and the frothy camouflage bubbles up. The larva possesses the ability to pump when it hatches out of the egg, deposited in the tissue of the plant by the ovipositor of the female Froghopper. The body of the larva produces the necessary material and the larva knows instinctively how to use it. When the sap dries up at the place which has been tapped, the larva moves further on and makes new shelters of froth, until after its fourth moult it is an adult Froghopper. It is then capable of flight and jumping and no longer needs camouflage.

Froth camouflaging Froghopper larva *(Philaenus spumarius)* with uncamouflaged larva (below). – *Magnified.*

On the banks of Southern European rivers during the mild May nights little dots of light can be seen in bushes and shrubs... this indicates that it is the breeding season of the Glow-Worm, a kind of beetle. Females and larvae possess organs which produce light; the males are winged and the females wingless. For a long time biologists believed that Glow-Worms contained light-producing bacteria, but today it is known that they themselves are responsible for the light. Their photogenic organs are plentifully supplied with tracheae and are able to manufacture light through the oxidation of luciferin by the enzyme luciferase. Thus the insects are enabled to indicate their presence to one another. However, their lights are not seen for long. Soon after laying their eggs — which are very light and from which hatch phosphorescent larvae — the female Glow-Worms become exhausted and die. Adult Glow-Worms feed on small quantities of pollen and nectar, but the larvae feed on molluscs, even sucking snails right out of their shells. — The Glow-Worm is native to Britain.

A luminous wingless female Glow-Worm *(Lampyris noctiluca)*, which leads a nocturnal life. — Magnified.

The wild rose bush often has mossy balls near the hips and haws known as "Robin's Pin-cushion".

The Gall-flies are a most interesting group of insects. Up to the 17th century it was still believed that galls were a diabolical manifestation, until in 1675 Malpighi observed an Oak Gall-fly laying its eggs and then followed the development of the gall. The particular Gall-fly which produces the Robin's Pin-cushion is the Rose Bedeguar Gall-fly *(Rhodites rosae)*. It is 4 mm long and lays its eggs only in the spring buds of the wild rose. One gall contains several larvae, each in a separate cell, and the galls are red and green in colour. They actually arise as a result of a chemical reaction brought about by the larvae themselves, the irritation they cause resulting in the production of cellular matter. The parasites remain in the galls until the next spring when the adult insects finally fly away after the gall withers. But the Robin's Pin-cushion contains not only the larvae of the Rose Bedeguar Gall-fly but also larvae of other species of parasites. The most familiar galls are those produced by Gall-flies which live on oak trees, and are known as oak apples. If these are examined early in the summer, the holes by which the adult Gall-flies escape are clearly visible.

A Leaf-cutting Bee has been at work here. The related *Megachile centuncularis* prefers rose leaves.

The world is full of building materials, if only one knows how to use them. Leaf-roller Weevils curl up whole leaves to form shelters for their offspring. But a Leaf-cutter Bee makes its brood even more comfortable; first it finds a hollow plant stem or a beetle burrow that seems right for a nest. Then it searches for a suitable leaf, cuts out an oval piece, and hurries away with it. Back at the nest-site the leaf is rolled up and pushed into the tiny hole, where it unrolls and papers the interior. The bee does this 15–40 times. Some of the Leaf-cutter Bees annoy gardeners by attacking the leaves of rose bushes, which they leave looking like pieces of surrealist fretwork. Osmia Bees line their brood chambers with red poppy petals, whilst the related Wool-carder Bees are upholsterers; they brush the fine down off fluffy stems and carry balls of it to their brood chambers, where they pull the down apart and use it to line the walls of the nest. All three bees fill the nest with nectar and pollen, lay their eggs and cover them up with the appropriate material. Everything is now ready for the next generation... so long as a parasitic bee has not managed to smuggle in its eggs in an unguarded moment.

Both bees and flowers depend upon each other for survival. Here, a bee has landed on the under lip of a flower of the sage *(Salvia)*, where the stamens are hidden behind the upper lip, which protects them from the wind. When it starts looking for nectar in the base of the flower its proboscis releases a mechanism, which bends down the stamen onto the back of the bee. The bee will then hasten to the next sage bloom, for each Hive Bee *(Apis mellifica)* is faithful to one species of flower: it recognises its particular flower by colour, shape and scent, and visits it when it has its richest store of nectar. Some bees act as scouts, and when they have found the right kind of bloom they perform a dance which indicates to the rest of the bees in the hive the exact direction, distance and size of the group of pollen-bearing plants they have discovered... and immediately the bees fly out in their thousands to collect the sweet nectar. For only 2 pounds of honey about 8 million different visits to flowers are needed — i. e. 8 million pollinations. But pollen is also collected as food for the larvae, for it is rich in albumen. To this end the bees begin even before the spring brushing off pollen from Hazel and Willow catkins, to comb it into the pollen sacs on the outside of their back legs during flight. When these are full they return to their hive and unload the pollen into the pollen cells.

Hive Bees must be continually on their guard against marauders in search of food. Therefore by the flight hole of every hive there are the sentinels that will sting to death any wasp that attempts to enter. Every hive bee has its own particular duties. The young bee hatches on the 21st day: from the 1st to the 3rd day it cleans the cell, also fanning the air, heating and carrying water. On the 4th day it begins to feed older larvae with pollen and honey. From the 6th to the 12th day the glands which produce royal jelly are fully developed and it becomes responsible for the care of the brood and the grooming of the queen. However, on the 13th day its wax-producing glands start to function and it becomes a cell builder. Its next duties are thickening nectar and pounding pollen, and from the 18th to the 20th day it in its turn becomes a sentinel. Only when its labour power has been used for 3 weeks by the "state" is it exposed to the risks of gathering pollen. At the most its life will last another 2 weeks. Nevertheless, as long as the queen is capable of reproducing, there will be another 2,000 to 3,000 new young workers a day to fill the gaps left by the older ones dying.

The Bee on watch, its wings whirring, warns the sentinels inside...

... and they leap upon the intruder: Wasp, Hornet, Death's Head Hawk or Wax Moth.

Many of the insects belonging to the order Hymenoptera, which includes wasps and parasitic flies, as well as ants and bees, perform a double service, both pollinating plants and getting rid of insect pests. Many of them produce larvae which require the flesh of other insects to survive and develop. While the social wasps feed their larvae with chewed broth, the Solitary Wasps line their cells which will contain larvae with stores of meat. Often a supply of fresh meat for the larvae which will hatch later is assured simply by paralysing or crippling the prey — by injecting poison into the nervous system. The different species of wasp produce poison which affects particular insects, so that one species is only capable of drugging caterpillars and others only spiders, bees, flies or grasshoppers. Some of the Spider or Pompilid Wasps, having paralysed a spider, actually lay their eggs on top of it, then hide it in a burrow. The Ichneumon-flies, however, do not even build a burrow, for they lay their 1 to 100 eggs — depending on the species — directly in a living caterpillar with their long ovipositors. The parasitic grubs then feed inside the caterpillar, or other larvae, and although the host seems perfectly healthy and continues to grow, it can never become a butterfly or moth. Some species of Ichneumon-fly lay their eggs inside the eggs of other insects before they hatch. Others pierce through the bark of trees to lay their eggs inside larvae feeding on wood. There are occasions where eggs are laid in a caterpillar which already has eggs inside it, and the larvae of the second species of Ichneumon exterminate those of the first.

← Ichneumon-fly *(Enicospilus merdarius)* laying eggs in caterpillar of a moth.

1 and 2: a Sand Wasp *(Ammophila pubescens)* carrying and burying drugged caterpillar.
3: Parasitic Pompilid Wasp *(Ceropales maculatus)* pursuing another Pompilid Wasp *(Pompilus plumbeus)* transporting its prey, a spider.
4: one of the Solitary Wasps *(Eumenes pendunculatus)* stocking its egg cell.
5 and 6: a Sand Wasp *(Sphex albisectus)* transporting and burying victim.

1

2

3

4

5

6

The young Ant Lion *(Myrmeleon formicarius)*, 10:1, a dangerous enemy to other insects, becomes a crepuscular insect like a dragonfly, 3:1.

Insects go through many metamorphoses, only a few of which are familiar to most people. However, students of insect behaviour today are already familiar with 750,000 species of insect and their changes in form. It is fascinating to realise that while very small versions of the adult insect emerge from the eggs of the Grasshoppers, Praying Mantis, Earwigs and Cockroaches, larvae which are completely different from the adult, with different organs and requiring different food, hatch from the eggs of the Butterflies, Beetles, Ants, Bees and Wasps, Mosquitos, Flies, Dragonflies and Ant Lions. This last insect is one of the most interesting of all. In the spring the larva hatches out of the egg: it is carnivorous, but not agile enough to hunt its prey, and instead it lays traps. If an ant should slip on the edge of the hollow in the sand where the Ant Lion is buried it is bombarded with sand which knocks it into the waiting jaws of the Ant Lion, whose digestive juices are immediately secreted onto the prey and the carcass is emptied of nourishment. Then in May the Ant Lion burrows down to a depth of 4 inches and pupates in a spherical cocoon covered with grains of sand. A month later the imago, or adult insect gnaws its way out of the cocoon, to emerge as a winged insect, greyish in colour, but resembling a dragonfly. It leads a crepuscular existence, flying from June until September.

Out of an egg of the Drone Fly *(Eristalis tenax)* . . . hatches this shapeless larva with the unusually long breathing tube, 4 : 1.

It is most interesting to observe the life of those larvae which use cunning or speed to overpower their victims either on land or in the water. Indeed they behave in such a determined manner that it is difficult to believe that they are only an as yet undeveloped form, hunting solely in order to accumulate, over a period of weeks, months or even years, the materials necessary to produce the imago or perfect insect. On the other hand, there are the larvae or maggots of the *Diptera* which do not show any activity but simply feed among the decaying matter where they live. In the very spot where the winged female laid her tiny egg, the larva of the Drone Fly spends one whole winter, living on the food it gets from manure and breathing through a long tube, which can be extended to many times its original length and which can be stretched to the surface in whatever position the larva is suspended. The larva grows and moults until its body has enough substance to form the imago. Only then does the insect pupate, the pupa being hard with a long tail: the larva disintegrates and hereditary factors build up the body of the adult insect out of the resultant mass. From the tiny visual points come huge facetted eyes, from the pouchy abdomen a body with limbs, and from mere folds of skin both wings and halteres. Soon the adult Drone Fly will emerge and feed among the flowers. Its name derives from the fact that it could be mistaken for a Drone Bee; but like all the true flies, it has only two wings.

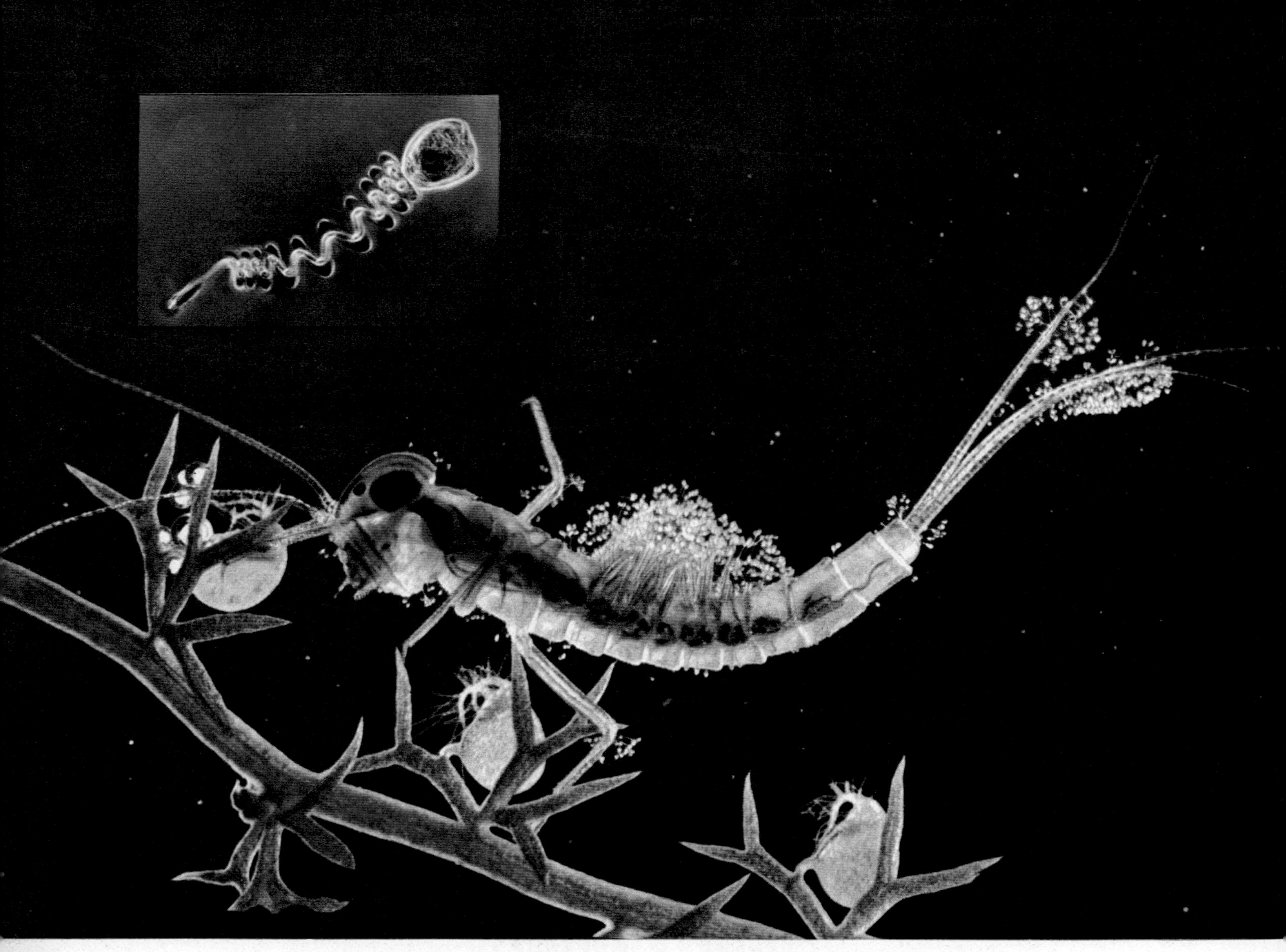

Some pond life: a Mayfly larva or nymph x 15. Inset: a Vorticel *(Vorticella nebulifera)* x 130.

For anyone equipped with a microscope and some degree of patience, a drop of pond water can reveal unsuspected worlds. And also in this minute universe, there are not only the hunters and the hunted, but also imposing bullies and furtive travellers. Thus the gills and caudal bristles of the Mayfly larva harbour hundreds of Vorticels. Unable to travel on their own, they use the Mayfly nymph to carry them to new environments, to water rich in the nutrients which they absorb. The carnivorous Utricularian, on which the Mayfly nymph is resting, lives on plankton and minute shrimps which it sucks in. The Mayfly nymph is also a voracious carnivore, and it needs to be, for it must complete some twenty moults during its aquatic life. But in those few precious hours as a completed aerial insect, eating would waste time. If the Mayfly's flight is so brief, that is because it lives only for love and vanity! Ephemeral insects, so soon to fly, so soon to disappear, for like the pages of a calendar, scarcely a day has passed before they are rejected, forgotten. As night ends the males flop exhausted onto the ground, and the females follow a little later, having deposited in the cool dawn water innumerable eggs, embryos which in three of four year's time will in their turn enjoy just a single day's flight.

Of the European spiders, only one *(Argyroneta aquatica)*, about ½ inch in length, spends its entire life under water.

The principle of the diving bell, used by Greek sponge fishermen 2,000 years ago, is the means by which the waterspider *Argyroneta* can exploit new hunting grounds under the water. Its only imitator in this sphere is the Tetragnath which propels a small raft about 2 ins square made out of grass stalks. But the little underwater structure of *Argyroneta* has only one purpose, to store oxygen. The spider comes to the surface to breathe, then dives again and starts spinning a fine silken web between aquatic plants. Again and again it comes to the surface for air, storing it between its legs and the fine hairs on its body. Then it plunges down to put the air bubbles under its web. Gradually the silver dome swells, and after about an hour the diving bell, now about ¾ inch in size, is completed. It is to this airchamber that the spider comes to eat the larvae and shrimps which it has caught. Like its relatives on land it must pre-digest its prey, and the gastric juices would otherwise be too diluted in water. It is also in this airchamber that the female receives the male, and here again that the young hatch out from the eggs. The *Argyroneta* only comes onto dry land when its fine body hairs are infected with minute parasitic fungae which tend to hinder the transport of air.

Bryozoans *(Bryozoae)*, magnified 40 times. Inside can be seen the Statoblasts, and at the left, some Vorticels.

It is nearly 300 years since the Dutchman Antony van Leeuwenhoek first discovered and described the world under the microscope. Nowadays a microscope is no longer a rare and expensive instrument, but few people think of using this "third eye" to explore the infinitely small and thus enlarge their own horizons. Even to amateurs a microscope can reveal astonishing beauty. Children will enjoy walks when they can look forward on their return to examining a jar of water taken from the lake, or a bundle of waterlily leaves. The slimy layer which covers these and other water plants conceals an undreamed-of spectacle, the flowering forth of the bryozoan plumes. Each one measures no more than $^1/_{10}{}^{th}$ inch but they form huge colonies, able to propagate in a number of different ways. Firstly by direct budding, or by detachment of the statoblasts. Secondly, each organism, being a hermaphrodite, can produce both eggs and sperm at the same time, the sperm being shed into the water and then drawn into a neighbouring polypide in which fertilisation occurs. Development continues until a small ciliated free-swimming larva is produced which is liberated into the water. Using these different methods of reproduction, Bryozoans have existed for more than 400 million years and the groups comprise some 3,000 species which have colonised salt and fresh water throughout the world.

Freshwater hydras *(Hydra vulgaris)* x 7. They can move by doing cart-wheels on their tentacles.

Freshwater hydras of ½–¾ inch can be seen with the naked eye, but only with a microscope can one really see their quite astonishing properties. Would you like to study their capacity to regenerate? You have only to cut them into small pieces, and from each piece larger than $^1/_6$th of a millimetre will develop a new polyp which before long will be catching little crustaceans with its tentacles. Under high magnification you can see something which does not appear in the photograph, namely the workings of the stingcells, weapons which are also possessed by jelly-fish, sea anemones and corals. Each tentacle bears thousands of small nematoblasts, each of which conceals a nematocyst, which is a long fine tube coiled in a spiral and armed with small hooks. At the slightest touch this spring uncoils, shooting out the stinging tube which is supplied with poison. Combined in hundreds, these poison darts paralyse the nervous system of the victim. Each barb is used only once, the animal continually making new ones. However, these formidable weapons meet their match occasionally, for it happens that some flatworms can ingest hydras without digesting the nematocysts. These borrowed weapons are passed through to their skin and are used in their own defence. Hydras occasionally reproduce sexually, but more usually by the asexual method of budding in which a daughter grows out of the parent body.

The front leg of the Water Boatman *(Dytiscus marginalis)* enlarged 120 times to show the complex adhesive organs.

Once one has succeeded in going beyond the limits of human sight, one will always use the microscope to investigate the functioning of various organisms. The microscopist, fishing for plankton, who suddenly comes upon a pair of Water Beetles overpowering a frog, sees the same thing as the photographer who took the photograph on page 124. But the real fascination of watching similar animal dramas day by day in dimensions of millimetres lies in details of perception rather than in the event itself. In every description of the Water Boatman we are told that the first three segments of the forelegs in the male have hairs and suckers with which it fixes itself onto the thoracic shield of the female during the act of pairing, which lasts for several days. The hairs on the legs also help in swimming. This gives the microscopist the inspiration to turn merely theoretical knowledge into a living experience both for himself and others. He catches a Water Beetle 1½ ins long and uses this as an object of study. When enlarged, the intricate structure of every single limb and their highly specialised mechanism can be recognized... and yet another example of adaptation to a certain way of life, in this case that of a fiercely carnivorous beetle, becomes revealed.

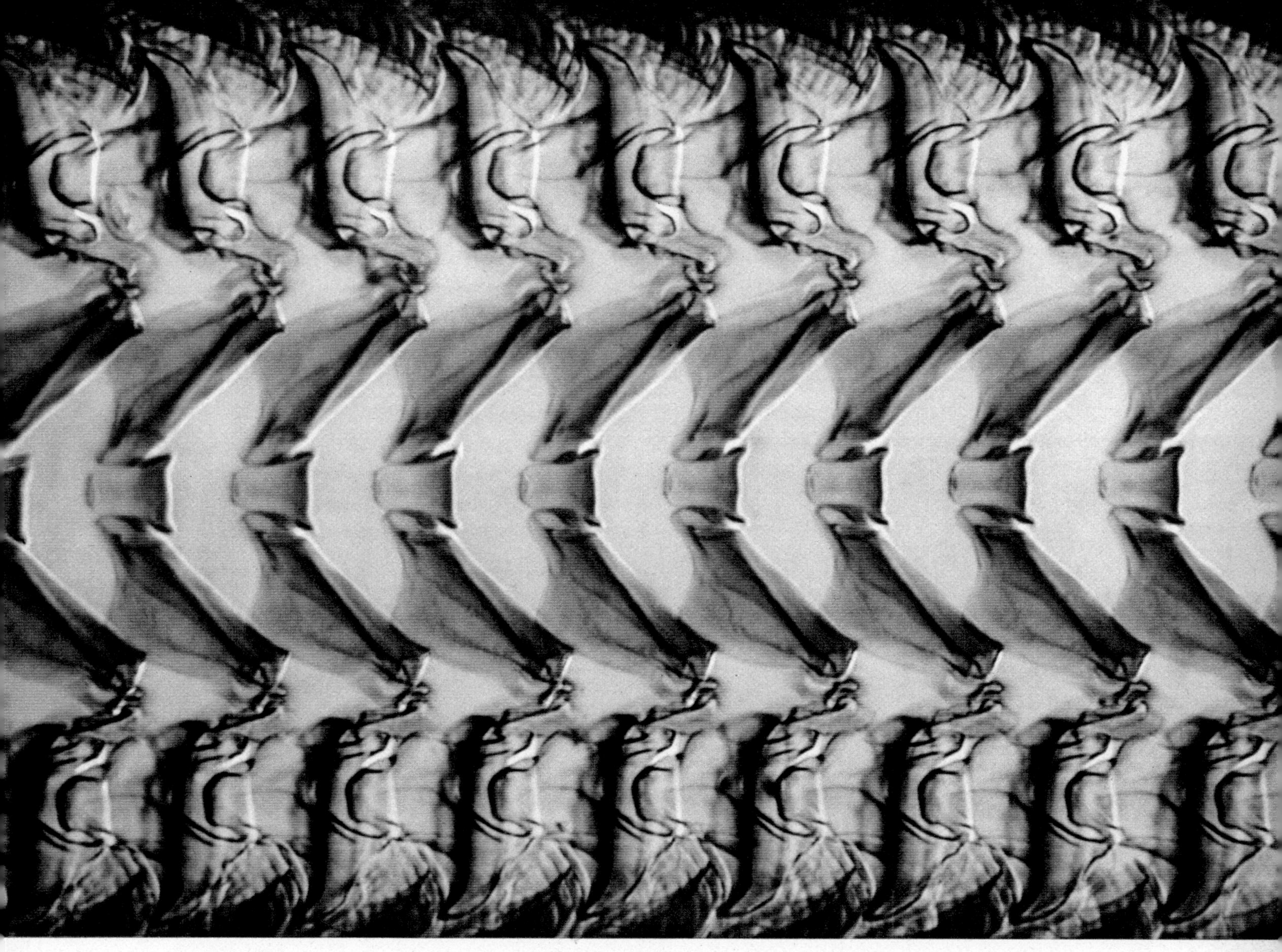

The zoological name for the tongue of both Snail *(picture)* and Shad is *radula*. Enlarged x 120, it resembles a file.

Even the familiar Snail can reveal surprising complexities under the microscope: for example, take a more detailed look at its tongue! This is like a file, which can grind down green leaves and even hard roots and tubers. The long horny ribbon, or frenulum, is supported by cartilage and is covered on top with tiny teeth, bent backwards, and differing in number and arrangement in each species of Snail. Therefore, to the malacologist, who studies molluscs, a Snail's tongue is often more interesting than its shell; for, in spite of the many varieties of this calciferous shelter, it is often difficult to distinguish between the 90,000 different Snail species of the world. Only a few of the air-breathing Snails, or *Pulmonata*, are found on the land, but in the sea there are an enormous number of inferobranchians which feed on algae or lie in wait for animal prey. The carnivorous gastropods have also developed other tools: chalk spines on their shells, with which they can prise open mussels, and an expanding proboscis with a boring device which clears a way to the actual mussel for the long tongue. In the Mediterranean *Conus*, the teeth on the tongue have even become poison fangs and the venom of the *Toxoglossae* from the Pacific ocean can even affect human beings.

The Spurge Hawk Moth *(Deilephila euphorbiae)* unrolls its long proboscis, with which it can reach the nectaries, pistils and stamens of nocturnal flowers. Its caterpillar feeds on the poisonous cypress spurge.

Insects have existed on the earth for as long as there have been forests. 300 million years ago there were giant dragonflies with wingspans of 1 yard whirring over the marshland. These giant forms of insect did not survive, though their fantastic shapes, their highly complex organs of sense and the implements which are part of their bodies have remained. Since the development of the microscope, countless new discoveries have been made. It is now possible to count the thousands and thousands of lenses of their many faceted eyes, their olfactory fossa (or organs of smell) and the nerve cells of their delicate feelers — all of which make their reactions to the world around them so very different from ours. For example, the unusual eyes of most moths and their extraordinarily sensitive taste and smell make them well adapted to a nocturnal life, for many can even detect the ultra-sonic sounds emitted by bats.

1 Fanned feelers of Cockchafer ♂, 2 Antennae of Black Arches Moth ♂, 3 Head and mandibles of the Mole Cricket, 4 Claw of Praying Mantis, 5 Antenna of Cockchafer ♀, 6 Jaws of Ant-Lion, 7 Eyes of Gad Fly, 8 Wing scales of Red Admiral, 9 Eyes of Salticid.

Although insects seem fragile, they are tough creatures. The carapaces, which in many kinds of beetles can support a weight a hundred times greater than that of the body, are hard and at the same time feather-weight and their pointed protuberances often serve as weapons. However, the limbs, feelers and wings which emerge through this armour are the most delicate things in nature. In the tarsus of a butterfly's slim legs there are organs of taste 2,400 times as sensitive as our tongues. The fan-like feelers of Cockchafer males contain over 50,000 sensory cells, and many moths possess antennae which can distinguish scents over distances of several miles, enabling them to find the female of the same species, whose wing scales produce a particular scent. As well as to scent, antennae are also sensitive to vibration. There are two kinds of mouths, the biting and sucking types, the biting type opening side to side and the sucking type up and down.

The Emperor Moth *(Eudia pavonia)* inhabits moorland and heath, where, if it escapes from owls, nightjars and bats, it lays its eggs on the stem of a plant. From the tiny eggs well over a hundred little hairy black caterpillars will hatch (1), which hungrily nibble the nearby leaves so as to increase their weight a thousand times in one month (2), if they manage to avoid birds' beaks for that length of time. After shedding their skin for the fourth time the survivors eat as much as they can and then find a nook or cranny, where they spin their cocoon using several hundreds yards of thread from special glands (3)... for these moths *(Saturniidae)* are related to the famous Chinese and Japanese moths, spinning cocoons which unrolled provide silk-threads up to 1½ miles long. Inside the cocoon, the pupa, which remains supple for a time, sheds the brightly coloured skin of the caterpillar which has now become useless (4). This skin is mainly black on hatching, but by three weeks later it is green, with black rings containing spots which are either gold or purple in colour.

1

2

3

4

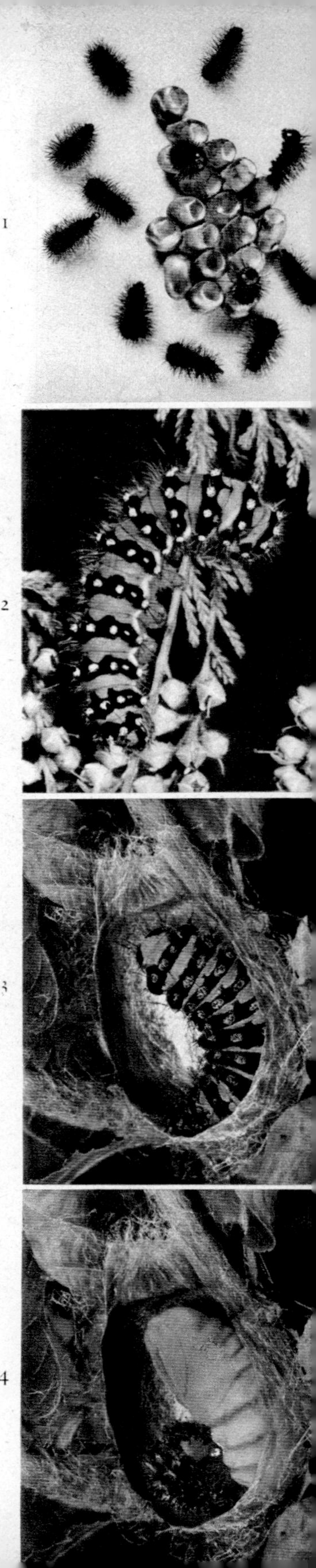

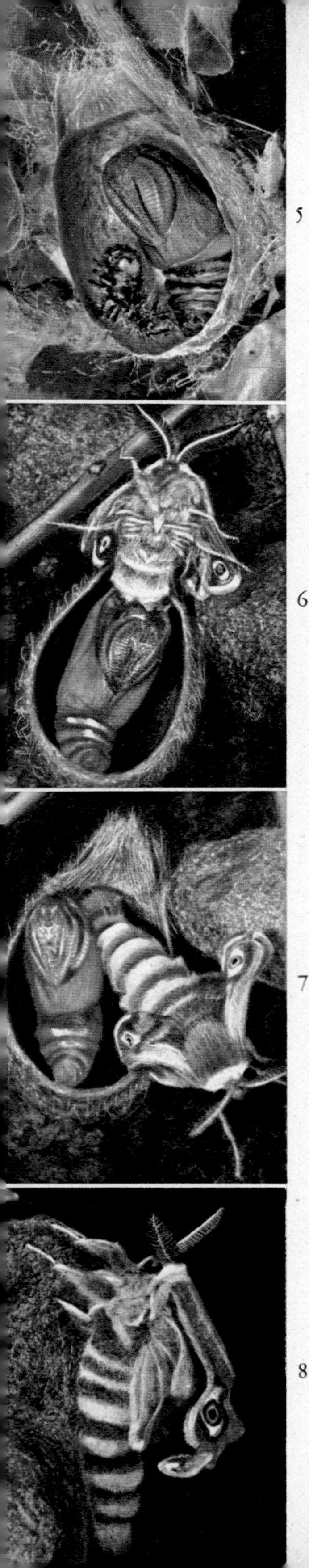

5

6

7

8

Later, when the pupa has become darker and harder (5), it is possible to recognise clearly the posterior rings, the feelers and the pouches of the wings of the future Moth. In this apparently dead chrysalis a further process in the life cycle is now completed during the long winter months: the transformation into an entirely different creature. At last in the spring the Moth struggles to emerge and finally succeeds in pushing its way through a slit in the head of the pupa (6); for these photographs the cocoon had to be opened, but normally the Moth must break it down with its digestive juices and carefully tear the woven fabric apart (7). While it rests after this great effort, blood is flowing into the short, crumpled wings (8); already it is possible to tell from the feelers that the new Moth is male. By sunset the wings will have stretched and hardened and the male Emperor Moth will immediately begin searching, by scent, for the female which is heavy with eggs. The adult moths cannot feed themselves, owing to their rudimentary proboscis, and therefore can live only for a few days.

Especially in the insect world, where so many enemies threaten life and even birth, every birth or transformation seems to be a miracle of survival. For example, take the case of the Elephant Hawk Moth caterpillar which has survived and has not been attacked by an enemy as a pupa. The moth which finally emerges from the cocoon in June hatches at the peak season for honeysuckle, phlox, jasmine and fuchsia. In the evening, the moths are attracted by the sweet scents of these flowers, and flying from bloom to bloom sip nectar from them. Relatives of the Elephant Hawk Moth are the rare Striped Hawks and huge Death's Head Hawks, which seem to be spreading Northwards. Since prehistoric times they have been searching for new places where they can feed and lay their eggs. There are also many migrant butterflies, such as the Clouded Yellows and Painted Ladies, which hatch in their millions in the heat of tropical Africa and make their way over the Mediterranean in clouds. Large numbers of them fly on over Central Europe as far as England, Scandinavia and even, occasionally, Iceland. In England the Painted Lady and the Red Admiral would die out if they were not thus reinforced every year. Frequently the caterpillars and pupae of both butterflies and moths fall victim to the Ichneumon-fly or other parasites and many are killed by the cold of Northern climates before completing their life cycle. In other words, it is almost essential for certain butterflies to lay their eggs South of the Alps if they are to assure the survival of the next generation of migrants. There are seven species of Vanessids in England, the Red Admiral, the Painted Lady, the Small Tortoiseshell, the Large Tortoiseshell, the Peacock, the Camberwell Beauty and the Comma. The Camberwell Beauty is interesting for it probably reaches Britain from Scandinavia whereas most other immigrants come through France or Belgium.

➤ Larva of the Ichneumon-fly in the opened pupa of a Painted Lady *(Vanessa cardui)*, 5:1. Its caterpillar (striped with yellow) was already the host of this parasitic Ichneumon larva, which gradually killed off the developing butterfly, and then continued to develop itself in the protective chrysalis.

➤ Elephant Hawk Moth *(Pergesa elpenor)* shortly after emerging from the pupa, 4:1. Its caterpillars are fond of vine leaves. It is protected from the cold night air by its silky covering. When the temperature drops it can heat itself up again to 93°, by whirring its wings.

In a sense insects such as butterflies, moths, bees, and bumblebees are responsible for maintaining the number of colourful flowers in fields and meadows, because they prefer flowers with bright colours. Throughout their history they have visited and pollinated such flowers, thus ensuring their survival. The life cycle of most butterflies and moths is dependent on the seasons when the flowers which they visit are in bloom. Only a few adult insects survive the winter, like the Peacock Butterfly and the Brimstone, which remain in a death-like state of torpor until spring when they lay their eggs. Most of the Lepidoptera pass the winter in an early stage of the life cycle. The Black Arches Moth, the Gipsy Moth and the Blue Butterflies spend the winter in the egg stage, development being slowed down temporarily; but the eggs of the Silver Washed Fritillary are laid earlier and the already developed caterpillar spends the winter still in the eggshell. The small Chequered Skipper in the colour plate spent the winter as a green caterpillar, which then pupated in its second summer, after a life of eleven months. The caterpillars of the Footman Moths, Wood Tiger Moth, Goat Moth and Swallowtail hibernate, and many caterpillars spend up to three winters under the snow of the Alps. The Wood Leopard Moth, for example, spends two winters as a caterpillar. But the vast majority of moths and butterflies spend the winter in the form of a pupa or chrysalis. There are, however, some species of butterflies and moths which produce two generations a year. In these, the influence of temperature on the life cycle can be observed: while the metamorphosis of the winter generation takes many months, the summer generation completes the same process in a few weeks.

➤ The yellow and buff Chequered Skipper *(Pamphila palaemon)* on Military Orchid, enlarged 6 times.

➤ These three brightly coloured Moth larvae spend the winter in the pupal stage:

1. The straw yellow, brown and black patterned caterpillar of the Chamomile Shark *(Cucullia chamomillae)* makes a hollow in the earth lined with spun silk.
2. The caterpillar of the Alder Moth *(Acronicta alni)*, black with bright yellow markings, pupates in a cocoon which it spins in decayed wood.
3. The caterpillar of the Sycamore Moth *(Acronicta aceris)*, which is covered with yellowish-red tufts of hairs, spins its cocoon beneath the bark of the tree where it feeds.

Pine Processionary Caterpillars *(Thaumatopoea pityocampa)*. They are harmful pests in coniferous forests.

The Pine Processionary Caterpillars have an inherited instinct to follow one behind the other. The inconspicuous grey-brown female Moths lay huge numbers of tiny eggs on pine twigs and there the caterpillars spin themselves a large common nest after hatching. In the evenings they emerge in a long column and wind through the branches in order to feed on the succulent pine needles. During this march the leader spins an endless thread out of its body fat which is constantly being strengthened by the other caterpillars, so that later on they can find their way back to the nest by following the thread. If after a time the enormous number of caterpillars on one young pine tree have reduced the amount of food to nil — then the long procession wends its way over the forest floor to search for another tree. The biggest of these chains of caterpillars which has so far been recorded was 13 yards long and numbered 300 caterpillars. When a whole forest of pine trees is attacked by these pests, it is important to deal with them rapidly. However, those who help by tearing down the woven nests from the trees must take care: the poisonous caterpillar hairs easily fall out and can do great harm especially to eyes and mucous membranes — the moth itself is harmless.

Female White-toothed Shrew *(Crocidura leucodon)* with train of young. It is an animal protected by law.

If danger threatens a family of White-toothed Shrews, the mother persuades one of the young to hold fast to her rump with its mouth. The other young all follow one behind the other and if one should get lost on the way, the mother will fetch it and carry it in her mouth. The female looks after the young until they are six weeks old and soon after this she bears another litter, usually of five young. White-toothed Shrews feed mainly on insects, but will also kill Field Mice on occasions. If they are attacked by an enemy, they let off a most unpleasant smell of musk which deters the attacker. Because of this smell the White-toothed Shrews found in Europe are also sometimes known as Musk-shrews. White-toothed Shrews vary in colour from slate-grey to dark brown above and yellowish-white below, with a clear division between the two. The adults are about 4 ins long, including a tail 1½ ins long. Whereas the common European Shrews have red tips to their teeth, the White toothed Shrews have white tips. They are not found in Great Britain or in Northern Europe, but inhabit fields and gardens in Southern and Central Europe. One race is confined to the Isles of Scilly off Cornwall and another to the Channel Isles.

View of the nest of the Yellow-necked Field Mouse *(Apodemus flavicollis)*, the largest European mouse.

A female Mouse can rear up to 30 offspring in a year, in 6 litters, and these in their turn are already sexually mature 8 to 10 weeks later. If it were not for the enormous number of enemies, carnivorous mammals, snakes and birds, which prey on them, the earth would be teeming with mice. The Yellow-necked Field Mouse, which is distinguished by the yellow spot on the throat, feeds on dandelion leaves and buds, birds' eggs and larvae, berries and roots. In the autumn it nibbles at beechmast and hazelnuts and stores many of these in the nest for the winter. In early March the first young of the year are born. While being suckled they attach themselves so firmly to their mother's nipples that they can be carried along if flight is necessary. If one of them should fall off it squeaks softly and is immediately picked up. After one week hairs begin to sprout on the young mice, and during the second week both eyes and ears open. After this if there is danger the mother carries her young away in her mouth, at which the young become stiff as if in a trance, which makes carrying them easier. They relax when they are dropped again. The Yellow-necked Field Mouse is a close relative of the Wood Mouse and both are found in Central Europe northwards to Scandinavia and from the British Isles East to Asia.

Female Common Hamster *(Cricetus cricetus)*. The numbers of these rodents have decreased in Western Europe recently.

The intensive cultivation of the soil has caused Hamsters to decrease in number, especially as they are solitary animals and do not like sharing the same territory. They live on the plains and especially in cornfields, in Eastern Europe and Central Germany, and need deep soil, without much moisture, where they can dig their extensive burrows. In these they store up plentiful supplies of food for the winter. They gather ears of wheat, peas and beans in their enormous cheek pouches and carry them to the burrow, often collecting over a hundred pounds in weight. During their hibernation they wake every three weeks to eat. In spring and summer, they sometimes eat molluscs, insects, or even young birds or mice. If they are threatened by a Marten or Buzzard, they adopt a defensive position and frequently escape; if a Weasel sneaks into their burrow, they fight back fiercely. The female Hamster bears 6 to 18 young after 20 days' gestation. At the slightest hint of danger, she hides the naked babies in her cheek pouches. As soon as the young play outside the burrow, they must learn to escape their enemies and take refuge in one of the many emergency tunnels, for even their own father may try to catch them. At 3 weeks old they become independent and have to dig their own quarters. The adults are 12½ ins long, including 2½ ins tail.

Not until the sun sets do the dormice come out to search for food among the nearby trees and bushes. Since they feed mainly on fruit, their food changes with the seasons. They are the first to harvest currants and cherries, pears and grapes, acorns, chestnuts and other nuts, moving from branch to branch and making good use of their long tails, which are particularly well adapted to climbing. To pick a ripe fruit, they will even make their way to the tips of branches, often hanging head down from their hind feet. As soon as dawn breaks they return to the nest, which may be in an old woodpecker's hole, a nesting box or an old bird's or squirrel's nest, which they have covered with a mossy roof. The Garden Dormouse, which is unknown in Britain, may build its own round nest in a tree, but it sometimes eats the eggs and even the young of birds and afterwards uses their nest. 3 to 7 young are born at a time. The Fat Dormouse, eaten as a delicacy in Roman times, has been introduced to a small area of Buckinghamshire and Hertfordshire.

Dormice hibernate for seven months of the year, consuming their body fat and occasionally nibbling at stored nuts.

➤ Garden Dormouse *(Eliomys quercinus)*, length 5 to 6 ins. Tail 4 ins. It lives in Western and Central Europe.

➤ Edible or Fat Dormouse *(Glis glis)*, length 6¾ to 8¼ ins. Tail 5½ ins. It is common in Central and Southern Europe.

By the twentieth day of their life young Squirrels venture out of the nest. At this time their small eyes are still hidden under a fold of skin; nevertheless, they have much contact with their surroundings, although they cannot see them: with a number of sensitive tactile hairs — over 60 — they find the safest route, hooking their sharply curving claws into cracks in bark. If they can go no further, they call until their mother carries them back into the drey in the branches of a tree. This is made out of dried twigs woven together and lined with hairs and moss. Later when the young can see they rush down the trunk, head downwards, and start learning to leap from one branch to another. The adult Squirrels can leap several yards, propelling themselves with their strong back legs. They also jump to the ground from great heights. After the first litter the female may bear one or more further litters later in the year. The main food of Red Squirrels is fir cones which they hold between their paws, gnawing off the scales and eating the seeds. They will also eat the buds from the growing trees, as well as acorns, beechmast and fungi.

Young Grey Squirrels *(Sciurus carolinensis)*. These Northern American relatives of the European Red Squirrels were introduced into England in about 1900 and have to a large extent supplanted the Red Squirrels.

Young Woodpeckers are already experts in the art of climbing vertically a few days after hatching. They do not hop about helplessly on the wood shavings in their huge nest hole for long, but soon climb up the inner walls of the nest and stretch their beaks out to take the ant larvae from the parent birds. At the end of May the young Black Woodpeckers leave the nest and cling to the trunk of the tree in which the nesting hole is excavated, sometimes as much as 30 feet up. They have strong toes and tails which support them during climbing and when their beaks harden they will be able to dig into the bark for the insects which have bored into the tree. Woodpeckers can also pick insects out of the cracks in the bark of trees with their sticky tongues covered with tiny barbs. Black Woodpeckers inhabit coniferous woods and sometimes beech woods in Northern regions. They are to be heard drumming occasionally and make a very loud noise. Their plumage is black all over and the male has a crimson crown and the female crimson on the back of the head. They fly heavily, swooping gently. They usually call while flying, "choc-choc-choc", rather like the Green Woodpecker.

Young Black Woodpeckers *(Dryocopus martius)*. Adults 18 ins long and Europe's largest Woodpecker. They are mostly resident throughout Europe but not in Britain, the Iberian peninsular or Italy.

In the summer Great Spotted Woodpeckers feed on insect larvae, but in autumn they add nuts to their diet. They can even break the hard shell of the hazel nut: if it will not open when banged against the trunk of a tree, the Woodpecker steadies it in some crack of the bark and splits it. But later when there are no nuts left and the wood-boring larvae mostly hibernate deep inside the trees, Woodpeckers gather fir-cones and pick out the little seeds from the scales. The Green Woodpeckers however, which spend much of their time hopping about on the ground, frequently break into ants' nests, which is the only harm they do, for where there are Woodpeckers in a forest they help in controlling the many insect pests which destroy the trees from within. During the breeding season the Great Spotted Woodpecker often bores more than one nesting hole in a deciduous tree or sometimes a pine tree, thus providing a convenient nesting hole for some other bird.

Great Spotted Woodpecker *(Dendrocopus major)*, resident throughout most of Europe, except Ireland and the extreme North. 9 ins long, weight 3 oz. Both parents incubate for 16 days. The 4-7 young fly at 3 weeks.

The male Hoopoe is bringing a delicacy for the female, which is brooding in a Spanish nesting box, made from the wood of the Cork oak. The entrance hole is too narrow for Martens or Weasels, but if any enemy should threaten the young birds, the female will secrete an evil smelling substance from the coccygeal gland and shoot it together with the faeces at any intruder. In order to collect food the male hops about on the ground in parks or gardens or open woodland looking for beetles, larvae, maggots and young lizards, and sometimes even pulling mole crickets out of their holes, nodding its head as it does so. When is catches some insect which it intends to eat itself, it throws it up in the air and catches it in its curved beak, so that it goes straight down into its gullet. The Hoopoe has a buffish pink plumage, wings and tail barred with black and white. The belly and rump are white and the legs are grey. The prominent crest can be spread out like a fan and, in flight, the Hoopoe often resembles a huge butterfly, particularly in its wing movement.

Hoopoe *(Upupa epops)*, summer visitor to Central and Southern Europe, wintering in tropical Africa. 12 ins long, weighs 2½ oz. Female broods for 16 days and the 5 to 8 young fly at 27 days.

The azure-blue Roller used to be seen in many European forests in summer up to a hundred years ago, but collectors of fine plumage gradually reduced their numbers. Only in the South and East of Europe do pairs of Rollers still return to their old territory in May at the beginning of the breeding season. At this time the male flies higher and higher over the treetops and tumbles downwards from great hights. The brilliant blue colouring of the Rollers is produced on the same principle as the blue of the sky. There is no actual blue pigment in their feathers — nor in any birds' feathers, or beetles' or butterflies' wings. There are fine translucent air-filled cells on top of a black ground, and these cells form a layer which disperses the blue light. If the pigmentation is yellow the colour we see is green and where the air cells overlap like scales, shimmering ultramarines and violets are produced. — The Whinchat, on the other hand, is brownish in colour, with darker streaks above. It is buff below and has a prominent eye stripe, white in the cock and buffish in the hen, and white chin and wing patches. It lives in open country and on derelict building sites, railway cuttings, marshland, commons and heathland, with bushes or bracken. It nests in rough grassland, often by a bush or shrub. It often perches conspicuously on a low bush, and catches flies during flight.

➢ A Roller *(Coracias garrulus)* alighting at the nesting hole, photographed from high up in a neighbouring tree. 12 ins long. Today only breeds in Southern and Eastern England, where it is a summer visitor; it winters south of the Equator in Africa.

⅄ The Whinchat *(Saxicola rubetra)* brakes during flight with wings spread out and fanned tail, before landing. It is a summer visitor from Africa to most of Europe except for Spain and Portugal, Italy and Greece. It is 5 ins long, with a short tail. Breeds throughout almost the whole of the British Isles, especially common in Wales.

There are about 150 species in the cuckoo order throughout the world; many of them incubate their eggs and rear their young, but all the 42 typical cuckoos of the Old World rely on other birds to do this. There are two European species of cuckoo, the European Cuckoo and the Great Spotted Cuckoo *(Clamator glandarius)* and both are brood parasitic in nesting: that is to say, they deposit their eggs in the nest of one particular species of song bird. Sometimes a single female cuckoo will lay up to 20 eggs during the 40 days when it is capable of laying, and those often resemble the eggs of the birds whose nests have been invaded. After hatching, the blind, naked young cuckoo instinctively pushes every object with which it comes in contact, whether egg or young bird, out of the nest with its strong wing stumps — so that only the huge gape of its beak remains to stimulate the foster-parent birds to bring food. The cuckoo is a summer visitor to the whole of Europe except for Iceland and the Faeroes, and usually migrates to Africa in August. The Great Spotted Cuckoo, on the other hand, only visits Spain and Portugal in the summer, and has wandered to Southern Europe, Finland and Britain. This particular cuckoo frequently lays its eggs in the nests of the Azure-winged Magpie *(page 51)*, often putting several eggs in one nest. It also lays in the nests of other members of the crow family. Another species of cuckoo, the North American Yellow Billed Cuckoo, has been known to wander North to Iceland and to Britain, France, Belgium, Denmark and South to Italy. It has a most unusual call, rapid and throaty. The European Cuckoo inhabits many different kinds of terrain, from open country and sand dunes to woodland and gardens. Besides the well-known notes it produces in the spring, it also makes a number of coughing and choking noises if excited and the female utters a bubbling trill.

Part of the development of a young cuckoo parasitic on the Dunnock or Hedge Sparrow *(Prunella modularis)*. The cuckoo female *(Cuculus canorus)* lays its eggs singly. According to where she was brought up herself, she entrusts her own offspring to Pipits, Wagtails, Warblers or Robins.

The Jay is almost always found in or near trees in both deciduous and coniferous woods, and is constantly on the watch, uttering its raucous alarm call at the approach of an enemy and thus warning all the other creatures of the forest. In the spring, however, it creeps stealthily through the branches to investigate the nests of other small birds. At the first chance it will seize a fledgling and make off with it towards its own nest. However, Jays are useful to foresters in keeping down insect pests and in spreading seeds. When the acorns and beech nuts ripen, they become vegetarians and often pick more than they can eat, burying the surplus. Such stores frequently remain buried, because the Jays rarely find them again. The noisy Jays are often to be seen in small parties, but when the acorns are ripe, they appear in large numbers on the oak trees. They fly weakly and heavily and have white patches on their rumps. The adults are pinkish-brown in colour, with black and blue bars and white patches on the wings. The feathers on the crown are streaked with black and white and sometimes have the appearance of a crest.

Jay *(Garrulus glandarius)* with offspring, which will leave the nest on the 20th day of life. Jays are resident all over Europe except the far North and parts of Scotland.

The song of the Turtle Dove is a soft purr which sounds almost sleepy. At breeding time the adult birds build their nest of twigs and both male and female brood for 17 days. When the two young hatch, the cells of the walls of the parents' crops multiply and produce a milky substance which collects in the side pouches of the crop. This nourishing broth or "pigeon's milk" is rich in albumen and fat and is poured straight into the gullets of the young birds, which flourish on this diet. By the fourth day, when the production of this substance diminishes, the young are able to digest softened seeds. The adult wild doves collect not only seeds and berries but also caterpillars and other larvae and pupae; even acorns can be softened in the crop, broken down in the antestomach, which is well supplied with glands, and then pulverized in the gizzard. Although we are used to seeing the more gentle domestic doves or pigeons, there are 400 wild species, and in the course of several thousand years human beings have bred another 100 domestic forms, from the decorative Tumblers and Fantails to the useful Carrier Pigeons which can carry messages at 95 mph.

The Turtle Dove *(Streptopelia turtur)*, 12 ins long, nests in small woods and thickets. It is a summer visitor, except to Scotland, Ireland and Scandinavia.

A Hen Harrier *(Circus cyaneus* ♀), with its wingspan of 45 ins, shades its young from the midday sun. But one of the chicks has left the shelter.

The Hen Harrier, a member of the Hawk family, is becoming rarer and rarer in the cornfields where more and more machinery is being used, but many still breed on moorland, in swamps and thickets. These birds of prey normally scavenge from the lakes near which they breed. As soon as the Harrier pair have prepared a nesting place by clearing a large area among reeds or on the ground, the brown female lays her eggs and starts to brood; meanwhile the all-grey male forages for food for his mate and sometimes drops it to her in flight. When, four weeks later, the five young are hatched, the male continues to hunt and brings back small birds, frogs and sick fish, drowned fledglings, water rats and sometimes even an old duck which could not escape. Later the female also forages and soon, one after the other, the young become independent. Hen Harriers are found in Europe, from Portugal to the Baltic Sea, and in Scandinavia except for the extreme North. They are also found in Northern Scotland and Ireland and have been known to breed in England and Denmark and to wander to Iceland. They produce both a high chattering sound and a long drawn out wail. They fly gracefully and buoyantly, sometimes gliding with wings slanting slightly upwards, frequently hovering. Female Hen Harriers are larger than the males.

The female Peregrine *(Falco peregrinus)* has seen the photographer and stands erect and staring, surrounded by the young.

Peregrine Falcons are habitually cliff breeders, laying their eggs on mountain crags, but when they occasionally nest in trees they use another bird's old eyrie. Both parent birds take part in the incubation of the eggs. They prey on many kinds of birds, especially on pigeons, ducks, grouse and waders; and, though seldom in Britain, also on mice and young hares and whatever else they can see in fields or meadows. The female, which is larger and much stronger than the male, swoops down almost perpendicularly onto her prey at tremendous speed, to break its neck with one blow of her sharp beak. Young falcons are taken on hunting flights by the parent birds from the age of 1 ½ months, gradually progressing from catching their food in flight to plunging downwards in order to seize it. The adult Peregrine is slate-grey above and has buffish underparts narrowly barred with black. The young are dark brown above, with streaked underparts. The Peregrine is a partial migrant to almost all Europe, except for Iceland, the Faeroes and part of Northern Greece. In flight it frequently soars and performs acrobatic feats. But, characteristically, it flies with swift wing beats, before gliding. The wingspan of the female is 4 ft. The unusual dark streaks at the side of the face resembling a moustache distinguish it from all other birds of prey except the Hobby.

A Barn Owl *(Tyto alba)* returning from a successful hunting expedition. It has caught and killed a pregnant vole.

When twilight comes and the diurnal birds of prey finish hunting, the owls emerge for their nocturnal hunting expeditions on the look-out for rodents. Many owls fly over forest and swamp on almost silent wings; the Barn Owl, however, hunts over fields on the outskirts of villages and inhabits old buildings, where it also breeds. It will even breed in clock towers or belfries. Its eerie, drawn-out, shrieking cry has for a long time given atmosphere to ghost stories. At breeding time it also produces snoring, yelping and hissing noises, and clicks its beak. It does not build a nest, but lays its white eggs at long intervals on the bare ground or boards in old buildings and occasionally on cliffs. Immediately the first egg is laid the female begins to brood and is meanwhile fed with mice or other small rodents by the male. The first young hatches after one month and the first down begins to appear on the third day. By the sixth day the young Barn Owl manages to devour a whole mouse, including skin and hair, later regurgitating the coat as a pellet. The adult Barn Owl has long feathered legs and a white face. It is pale golden-buff above, finely mottled grey, with white underparts, not always speckled. It is mainly a resident bird in the British Isles, Southern and Central Europe, with the exception of Sardinia and Southern Greece. However, it has been known to wander north to Finland.

Four young Barn Owls wait for their food. 93 per cent of this consists of rodents which are a pest to farmers.

It may be six weeks after the laying of the first egg before the youngest of the brood can fly. The Barn Owl's flight is slow and flapping, similar to the other large owls. Sometimes the male must hunt in the daytime as well as at night in order to obtain enough food for the young. So as to protect its eyes from the daylight it narrows the pupils considerably and half closes the upper lids. The world to the owl looks very different from that seen by a human being, because the arrangement of rods (56,000 per mm^2) and cones, of which there are much fewer, is adapted to a nocturnal existence, and probably gives an effect like that of a black and white film. The enormous eyes make up a third of the weight of the head and their form is such that the owl has binocular vision, enabling it to pick out even a tiny rodent, and its ears are so sensitive that it can hear soft rustling noises from a height of 10 to 18 ft. The already keen hearing of the owl is made even more acute by tufts of feathers on the ears which absorb sound. It also has tactile bristles on the beak, known as *vibrissae*, which are extremely sensitive to the slightest vibration. The Barn Owl has a huge white facial disc, which is formed by the flattening of the face. This is common to all owls; they can also turn their huge heads through 270 degrees, to see behind.

The Short-eared Owl *(Asio flammeus)* is the only owl which builds its nest on the ground on marshland.

The Short-eared Owl nests as far north as Iceland, where the white Snowy Owls hunt for lemmings. However, it is a partial migrant and sometimes families of 12 owls set off all together towards the south. They breed in a few parts of the British Isles, in Spain and Greece. They hunt by day as well as night and they are often to be seen flying and gliding over marshland and moors, closely resembling the buzzard. They fly swiftly and soundlessly, their long barred wings having, at the ends, horny feathers which dampen every sound. They usually perch on the ground and when there are many rodents about they hunt in parties: a fieldmouse does not realise that its enemy is above it until it is caught. If, however, a victim does try to defend itself, then the thick plumage on the owl's legs protects it from being bitten. It has been known for an Eagle Owl to plunge down upon a Short-eared Owl and kill it. But now this large Eagle Owl, 26 to 28 ins long, with a wingspan of 5 ft, which can even kill Martens and Hawks, has been driven to retreat into protected mountain forests, and so many have been shot that only a few of these magnificent birds still survive.

This Little Owl *(Athene noctua)* has found a warm nesting place in the straw; usually it nests in holes in trees.

The Little Owl often appears at night near human habitations, usually in farming country. It utters a shrill plaintive cry, which for a long time terrified country people. It is only 8 ½ ins long, very small in comparison with nearly all other owls, and squat, with longish, rounded wings and a short tail. It is now the most common nocturnal bird of prey in Europe, being introduced to England and a resident in the rest of Europe except for Iceland, Scotland, Ireland and Scandinavia. It is not, however, the smallest European owl. That position is held by the Pigmy Owl, which is only 6 ½ ins long. This tiny owl kills rodents and sometimes even bats, and also hunts small birds which it kills in flight. One of the most stately owls is the Great Grey or Lapland Owl, which inhabits the dense coniferous forests of Northern Scandinavia. The Tawny Owl, which is resident throughout most of Europe, is 15 ins long but has a wingspan of 1 yard. Its main prey consists of mice and voles, insects and a few small birds. It nests in holes in trees, the empty nests of large birds and sometimes in old rabbit burrows. During the daytime it is often surrounded by noisy scolding parties of small birds.

Emerging at twilight, Bats hunt in the night sky for
layers of skin stretched between the limbs and tail, like p
be used for climbing and hanging. The 60 million years
conquest of the air by mammals, are at once the archet
radar-contro

1 Greater Horseshoe Bat *(Rhinolophus ferrum-equinum)*, wir
↔ 8 ⅓ ins.—3 Lesser Horseshoe Bat *(Rhinolophus hipposid*
myotis), ↔ 16 ⅔ ins.—5 Late flying Serotine *(Eptesicus se*
length 1 ½ ins. ↔ 10 ½ ins.—There are 13 spe

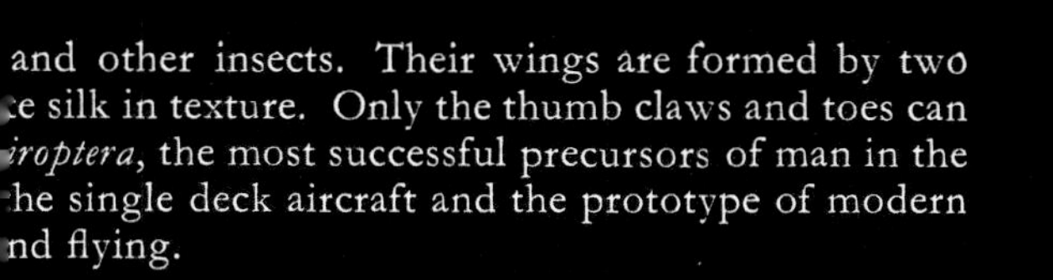

and other insects. Their wings are formed by two
:e silk in texture. Only the thumb claws and toes can
iroptera, the most successful precursors of man in the
he single deck aircraft and the prototype of modern
nd flying.

= ↔ 14½ ins.—2 Pipistrelle *(Pipistrellus pipistrellus),*
ngth 2½ ins, ↔ 9½ ins.—4 Mouse-eared Bat *(Myotis*
↔ 14½ ins.—6 Long-eared Bat *(Plecotus auritus),* ear
Bat in Great Britain, from 1½ to 3 ins. long.

2

3

5

6

The most unusual group of mammals is the Bats. They have many peculiarities besides their characteristic flight. In late autumn bats pair off and mate, emitting a musky scent. Soon after this they leave their usual sleeping places in lofts, church towers and hollow trees and look for caves and cellars. Often thousands of them gather together there and experience a sort of hibernation for the next 4 to 5 months. Oddly enough, during this period, as well as during their ordinary sleep, their temperature sinks with that of their surroundings; when the temperature sinks as far as 23° F they might freeze to death were it not for some powerful instinct or reflex action which wakes them at the last moment out of their deep lethargy. Thus hibernation is frequently interrupted by short spurts of activity. During this hibernation there is a pause in the process of gestation, which does not continue until the bats wake in the spring. Many pregnant female bats gather in one cave in early summer. In the act of birth each one hangs horizontally from a ledge and catches the newborn bat in the tail end of the flight membrane. Immediately after birth young bats cling to the belly of the mothers and seek the nipples. They are so firmly fixed there that they can be carried about in flight.

Serotine *(Eptesicus serotinus)* with its young, which will learn to fly at the end of the first month.

The human solutions of radar and sonar to the problems of night flying have increased our understanding of how bats manage to hunt in total darkness. Moving soundlessly through the air they keep their mouths open all the time, emitting ultra-sonic sounds which are inaudible to man. As they become more excited in the hunt for prey the pitch of these sounds rises from 30,000 to 100,000 vibrations per second (30 to 100 kilocycles). Telephone wires and even tiny mosquitoes in the line of flight reflect these sounds, which the bat registers with its complicated ear mechanism, distinguishing echoes from moving objects and from stationary obstacles. In the case of the Horseshoe Bats it is possible that the appendage of bare skin round the nostrils helps in the transmission and receiving of sounds. Up to 100 mosquitoes and a dozen Cockchafers are devoured every night. The number of places where bats can roost is becoming fewer, and ringing has revealed that in the autumn many bats roam for a hundred miles or more to find the peaceful caves around the slopes of the Alps. Long-eared Bats, whose ears are almost of body length, roost in towers, lofts and out-houses, hibernating in hollow trees, caves and cellars, and are found in large numbers throughout the British Isles.

Long-eared Bat *(Plecotus auritus)*. During roosting the actual ears are tucked away under the wings.

The Crested Porcupine *(Hystrix cristata)*, the Italian *istrice*, is up to 27 ins in length, with spines 16 ins long.

When the continents of Europe and Africa were still connected by land many animals came from the south to find a northern habitat. Fossils found in layers of soil 8 million years old show that whole herds of giraffe, zebra and antelope grazed on the ancient steppes of Greece until they were forced back again to their country of origin by the onset of the Ice Ages. But there was one African immigrant animal which remained in Sicily after the Mediterranean had completely cut off Europe from Africa: the curious porcupine — a unique feature of the fauna of Italy. Today this nocturnal hunter trots across the barren uncultivated land of Southern and Central Italy and only disappeared a short time ago from Tuscany and the Abruzzi. This phlegmatic rodent spends the day in a burrow in the earth which it digs itself and only sets out at night to forage for roots, plants and berries. If some other animal tries to attack it, it ruffles up its spiny armour, and seeks to awe its opponent by gnashing its teeth, stamping its feet and rattling its hollow tail spines. If this has no effect, it then rolls itself up into a ball like a hedgehog... if the enemy now tries to bite it, the long, easily shed spines at the rear will pierce deeply into the flesh of the attacker which will slink away with a swollen nose and may never attempt to attack a porcupine again.

An example of the Afro-European migration of animals: the European Genet *(Genetta genetta)*, which grows to 36 ins long.

Many North African animals once thronged across to Spain and Portugal over what are now the Straits of Gibraltar. Even African apes moved through France as far as Southern Germany where they remained for thousands of years, although today the only reminder of this ancient invasion is the presence of a colony of Barbary Apes on the Rock of Gibraltar. Yet some Civets probably reached Europe at the same time, and have managed to survive in the wild state. In Southern Spain the mongoose still hunts for snakes, and the long-tailed European Genet *(picture)* creeps through bushes and treetops of the Iberian Peninsula, hunting, for rodents and plundering bird's nests. Even in the South of France this animal can be found taking eggs from dovecots and hen houses... but while it may be observed doing so, it is extremely difficult to capture, for the Genet is more wily than the fox and more agile than the wild cat. It is not yet known *where* the male and female meet and pair after being attracted to one another by the smell emitted from the musk glands, or *how many* young the female later bears in some hidden cliff cave. Thus, even in our densely populated continent, an ancient beast of prey can still keep some secrets of its existence. Our domestic cat was no doubt just as mysterious to 11th century Europeans.

From her high perch the female Polecat *(Putorius putorius)* has discovered a mouse in the grass. One light spring...

Human beings have never objected to such animals as the Martens preying on small rodents — but they do object when game birds and hares are attacked. Add to this the fact that Martens have thick, shining pelts, and it becomes obvious why they have been hunted for such a long time. Where Martens and other carnivores have been exterminated the damage done by rodents to crops has increased enormously. It was not until the descendants of those Polecats, Stoats and Martens which had been driven to retreat into isolated places, again ventured to approach human habitation that the menace of rodent pests was once more reduced.

Polecats usually inhabit fields, moors and marshland, and sometimes in winter will enter hay-lofts. Pairing takes place in March and in May the young are born. During mating male and female Polecat stalk one another, hissing and shrieking. Afterwards the males go on their way and mark the boundary of their terrain with their distinctive jets of odour. A patch of water is usually included in the territory, for as well as preying on snakes, insects, worms, birds and rodents, Polecats sometimes vary their diet with a meal of fish, frogs and tadpoles. Polecats have been exterminated in most parts of the British Isles, except Wales.

. . . and the unsuspecting victim will never gnaw at roots again. Nearby the young polecats wait for their food.

After 2 months' gestation the female Polecat bears 3 to 7 nearly white, blind young in an old fox or badger earth, in a hole dug under a tree root, or even in an old hay loft. Before going on hunting expeditions she carefully covers the tiny young with grass; and 3 weeks later, when the young have their eyes open and their fur is darkening, she brings them their first small mouse. A fortnight later the young leave the burrow to play and soon they too will take part in the hunting expeditions, learning how to take a mouse by the scruff of its neck to avoid getting their noses bitten, and learning to scent the Stone Marten in the rabbits' burrow. Polecats even kill vipers; and if they come across a hedgehog they will attack it by first emitting a stench which renders the hedgehog incapable of action, and then rolling it over on its back without being pricked by the spines. The Polecat can also carry eggs clamped tightly under its chin to prevent breakage, and can paralyse frogs by biting through their spinal cords, after which the frogs remain fresh for days. The young are quite capable of emitting the characteristic stench of the adult Polecat. The adult is 21 to 24 ins long and has a thick pelt with fewer outer hairs through which the yellowish colour of the underpelt can be seen quite clearly.

Before the fox cubs are born in April, the vixen drives away the male as a precautionary measure and afterwards has to provide for the litter herself. Sometimes, if the male does not return, a vixen, desperate for food, will break into a hen house. Foxes are very wary and normally hunt at night, sniffing the ground to scent their prey, which varies from grass snake to mole, from wild fowl to hare and includes thousands of mice but very few fawns. They also add insects and berries to their diet and in winter feed on carrion. In July the cubs start to accompany the vixen on hunting expeditions or catch beetles and grasshoppers on their own. Where the larger carnivores have been exterminated, the lice in their own fur are the only natural enemies except for man, who menaces them with shooting, traps and poisoned bait. The fox-hunt is still a social event in many places, and in Britain some 230 packs of hounds take a yearly toll of about 13,000 foxes.

Young foxes *(Vulpes vulpes)*. There are usually 5 to 8 cubs to a litter, each weighing 4 oz. at birth. When they emerge from the earth in May, their fur is greyish-brown in colour; the adult measures 42 to 46 ins in length and is usually russet above and white below. Foxes are found throughout Europe, extending into Asia.

These Wolves *(Canis lupus)*, alarmed by the photographer's helicopter, show aggressiveness and fear.

There have been no Wolves in the British Isles since the 18th century, and they have also been exterminated in Denmark, Holland and Switzerland. Today they are mostly found in the Northern hemisphere, especially in Finland, in the nature reserves of the High Tatra which are rich in deer, as well as in the woods of Poland and Czechoslovakia. Since the end of the Second World War, however, the numbers of Wolves in Eastern Europe have once again increased, and more are now appearing in Italy and, recently, in France. They mostly inhabit forested country, and gather in packs in the winter, roaming great distances. Occasionally Wolves from Eastern Europe travel as far south as the Appenines and Pyrenees. In Spain the scourge becomes so great at times that 6,000 pesetas are offered to anyone killing a Wolf, and sometimes at night villagers have to defend their livestock against attacks by hungry Wolves. Later, in spring, when mating takes place, Wolves settle in one place for a time and the female bears 4 to 6 woolly cubs in a hole. These cannot follow the pack until their third month of life, and they remain with the same pack for 3 years until they are sexually mature. The adult Wolves are 58 to 70 ins long and resemble Alsatian dogs. As well as preying on mammals and birds, they will sometimes feed on soft fruits or carrion. They are well known for their melancholy howling by means of which a scattered pack can be gathered together again or the presence of some prey can be indicated.

A young Glutton or Wolverine *(Gulo gulo)* came to gnaw at the remains of an Elk, probably killed by the female *"järv"*.

Despite the many attempts made for over a century to get rid of the Wolverines of Norway and Sweden, the animals still survive and still go on their nightly hunting expeditions. In 1945–46 a bounty of 3,500 kroner was offered in Norway for each Wolverine killed. In winter these solitary Gluttons are responsible for mass slaughter on many herds of Reindeer, an animal which they can overpower and kill in an instant. After they have eaten their fill they drag the prey to a safe distance and finish it later, occasionally burying some of the meat in the snow. Wolverines are very agile and run well even in deep snow. They can also roll great distances, springing up on all four feet at once, and quickly escaping at the first signs of danger by rushing off into the snow of the tundra. They inhabit woodland and mountain regions of Norway, Sweden and Finland, Russia and parts of Asia, and also North America. On their nightly trips they kill mainly small rodents, Lemmings in particular, but also eat berries and birds. The 2 to 3 small young are greyish and are born in March or April in a hole in a tree or a rock crevice. Adult Wolverines are 33 to 38 ins long and their coat becomes darker in colour as they grow older and also in winter. Scandinavian superstition has portrayed the Wolverine as much more fierce and destructive than it actually is, and indeed even in the English press as late as 1888 there were stories of its ferocity — though of course it was not found in Britain.

A young Arctic Fox *(Alopex lagopus)* following a track by scent. The fur of the young is grey and woolly and in the winter it turns white.

There are only a few animals which live beyond the Arctic circle and have dark winter coats like the Wolverine. Those animals which have thick white fur in winter have two advantages, for not only does the coat act as a camouflage to prey and enemy alike but it also helps to maintain the body temperature at a constant level, since white, although absorbing little heat from the sun, gives out very little. The long hairs, densely packed together, act as a trap for warm air, and so warmth is stored up in the coat. In summer, when the Arctic Fox roams through crags and cliffs in the far North, including Iceland and the extreme North of Norway and Sweden, looking for sea birds' nests, it is a greyish-brown, thin creature with a narrow tail, but in the winter it has a thick coat, a bushy tail and hairy insulated soles. It does not look much like the fox which inhabits woodland further south. The Arctic Fox will eat almost any carrion thrown up on the coast, dead whales or fish and even skins snatched from trappers. Their chief food, however, consists of Lemmings, mice and birds, especially the Ptarmigan. Often several pairs of foxes come together to excavate the earth, which has up to 50 or 100 entrance holes. The 6 to 8 young are born in late May. Fur trappers in the far North are interested in a particular colour variation — known as the Blue Fox — which has a blue-grey fur the whole year round.

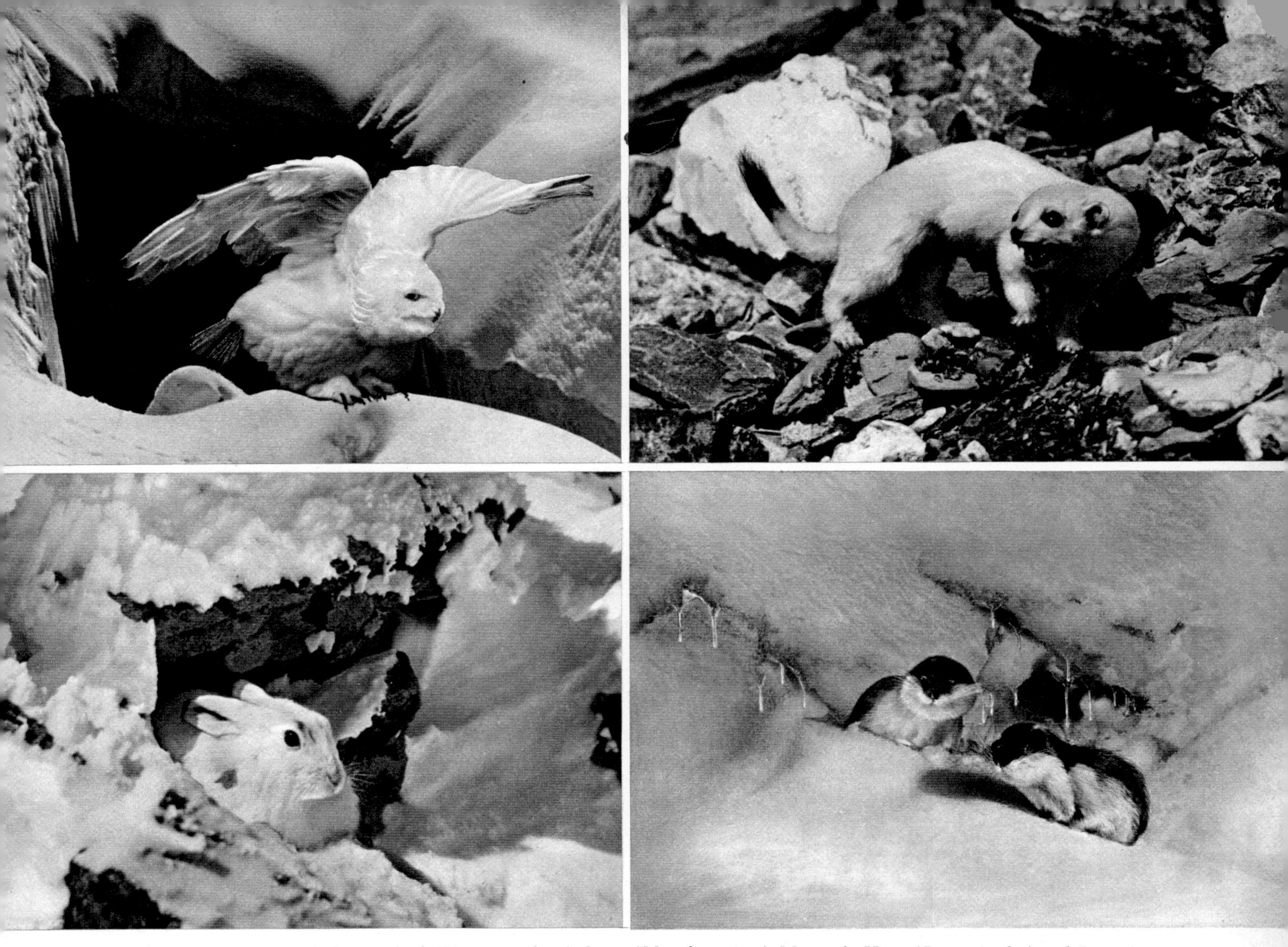

Hunter and hunted in the snow: Snowy Owl *(Nyctea scandiaca)*, Stoat *(Mustela erminea)*, Mountain Hare *(Lepus timidus)* and Lemmings *(Lemmus lemmus)*.

In the snows of the Arctic those vegetarians and carnivores which do not hibernate must fight every day for each mouthful of food. Soundlessly the gigantic Snowy Owl, 21 to 26 ins long, sweeps over the tundra of the Arctic, trying to make out Mountain Hares and Lemmings in the snow and on the rocks. When a Lemming emerges for a moment from the snow into the light, the Snowy Owl must catch and kill it; the Arctic Fox, however, can dig for one of these small sharp-toothed rodents whenever it comes across one of its entrance holes. Norway Lemmings dig down into the snow during the winter. They make passages underground and immediately take cover in these if they sight an enemy. Nevertheless, the Stoat is able to slip into these burrows and pursue the squeaking and hissing rodents through the maze of tunnels with their lichen floor and roof of snow. Both the Stoat and the Mountain or Blue Hare have produced great varieties of colour: in the extreme north they both remain white and thus well-camouflaged throughout the year. However, the Irish sub-species of Mountain Hare, *Lepus timidus hibernicus*, and the Central European Stoat still have their brown coats in the winter, while the Alpine Hare and the Stoat of the mountain forests change their colour with the seasons, so that hunter and hunted are camouflaged in the same way. The Stoat is brownish in colour during the summer, its fur becoming mottled for a short time and turning completely white in winter, except for the black-tipped tail.

Young Siberian Jays *(Perisoreus infaustus)* huddle together for warmth after their first hunting flight.

In thick northern pine forests and birch woods may be found Siberian Tits, Goldcrests and Waxwings, and still further north, above the tree line in mountainous regions and on cliffs, live the Redpolls and the dazzling white Snow Buntings. Their young are threatened by the fox-red Siberian Jay, collecting food for its own brood. In winter time these Jays flock to the outskirts of villages and have even been known to wander as far south as Czechoslovakia in years when there are not many cones on the pines and firs of the Northern forests. They are normally gay and perky birds and produce a cheerful "cook, cook" and a harsh "charr". Like the Jay itself, they also appear to imitate other birds, for example the cat-like cry of the Rough-legged Buzzard, the piping of the Starling, the trill of the Redpoll and high trill of the Waxwing. Siberian Jays may be seen clinging to the ends of branches of fir trees in order to reach the cones. They usually make their nests in pine trees, near the trunks. During the breeding season they become more timid and are nearly always silent. They are found in Northern Scandinavia, and the southern part of their range, where they are mostly resident, extends a short way into the range of the Jay. The Redpoll, on the other hand, which is found in the extreme north of Scandinavia, in North-Eastern Iceland and throughout the year in the British Isles, occurs as far south as the Italian Alps. This bird breeds in juniper trees, alders, willows or birches, often gregariously. It is 5 ins long and grey-brown, the male having a pink breast.

Closely huddled together, with ruffled-up plumage, a covey of Partridges *(Perdix perdix)* defies a winter snowstorm.

Of the game birds, the Willow Grouse, Ptarmigan, Capercaillie and Black Grouse spend the winter in the extreme north; the Partridge only extends to Southern Norway and to the north of the Gulf of Bothnia. It is resident throughout most of Central and Southern Europe except for Spain and Southern Portugal and Southern Greece, nor is it found on the Mediterranean islands. It makes a large nest-hollow under a hedge or in a cornfield and hatches up to twenty inconspicuous young, which, from the time they are born, can move at lightning speed to huddle together at the first sign of danger. Many other birds and mammals prey on the young Partridges while they are small, and as soon as they learn to fly in September, the marksmen are after them as well. As Partridges remain in the field, pasture or moorland where they hatch and always stay in their covey, it is easy for a retriever to chase and frighten them so that they fly up into the air in a party and two or three can be shot without difficulty. When Partridges are excited they produce a penetrating cry which they repeat at great speed. In many places in Europe these chicken-like birds have very nearly been wiped out by senseless shooting; and new stock, introduced to strengthen the old, has instead sometimes brought disease and thus killed off the surviving native Partridges. They have been introduced to North America, where they are called Hungarian Partridges. Partridges are very important to farmers for they kill and eat many pests — in fact 80 Colorado Beetles have been found in the crop of a single male Partridge.

All over the world the domestic fowl, descendant of the Southern Asiatic Banciva Hens, constantly lays eggs for man to eat. At the same time its wild relatives, the game birds, inhabit field and woodland, cliff and snow covered mountains. In the case of the domestic fowl, the male is brightly coloured and has the familiar red crest, while the female is a plain inconspicuous little bird; but among other species male and female are hardly differentiated. During the display ceremonial of the Ptarmigan both male and female have the unusual light and dark plumage, in irregular patches, that is a transition stage between the winter and summer plumage and resembles the spring landscape in the high hills. Both male and female Red-legged Partridge also have the same plumage, and a pair of these birds will remain together for many years. The male normally keeps watch, and takes no part in incubation; but he has been recorded as sitting on a separate clutch laid for him by the female. A hormone is released into the bloodstream of the female by the pituitary gland and has an effect similar to sedation. At the same time the blood vessels in a particular part of the skin of the belly become enlarged so that here the eggs can receive the benefit of the full body heat (104°). The Red-legged Partridge must brood for 25 days: however the longest incubation period is that of the King Penguin and the Condor which lasts for a full 2½ months. Like other game birds, the Red-legged Partridge sits very tightly when incubating and will allow a photographer to approach to within a few feet... and it is possible that even a fire would not make it desert its eggs.

➤ Brooding Red-legged Partridge *(Alectoris rufa)*. Now resident on the Iberian peninsula, in south and eastern England, Corsica and Northwestern Italy. They were introduced to England in 1770 and adapted themselves well to the thicker vegetation and damp climate. They are 13½ ins long and can run very fast. They inhabit sandy ground, stony wasteland, and sometimes marshland.

➤ Four stages in the transformation of the plumage of the Ptarmigan *(Lagopus mutus)*: 1. Winter plumage; 2. Beginning of the spring moult; 3. Plumage at the breeding season; 4. Autumn plumage, providing good camouflage. Underparts and wings always remain white, but in summer the lighter wings are kept well hidden and in winter the black tail tip is concealed beneath the camouflaging plumage. The Ptarmigan is found on high treeless mountain slopes in the Alps and Pyrenees, where it has existed since the Ice Ages. They are also found in Scotland, Norway and Iceland. The variety *Lagopus scotius* or Red Grouse is resident in Great Britain (except for the South East) and Ireland.

Anyone wishing to observe the display ceremonial of the Grouse or Tetraonidae, which include such birds as the Ptarmigan, the Capercaillie and the Hazel Hen, must set off during the very early morning in the early spring in order to hide in some convenient spot before the sun begins to rise. With wings whirring, the huge Capercaillie sweeps overhead to settle on the thickest branches of a nearby tree, where it begins the first stage of its display. This consists of very soft wooden clucking noises and a bell-like tinkling repeated in rapid succession until finally this stage ends with excited "whispering", and grinding noises, made partly by the wings. By now several hens will have begun to gather in the clearing by the tree and the male bird sweeps down onto the ground where it performs an elaborate tripping dance. The display and the pairing may be over before the morning mists clear, and the male may dance once again in a sort of frenzy after the act of pairing. The males very seldom fight, unless the number of females has decreased through the ravages of foxes and martens. However, in the case of the Black Grouse, which are much smaller, a sort of tournament takes place at the "lek" or display area, where the males vie with one another, springing high into the air, in order to deter each other and to impress the females. During the display the males produce loud rapid gobbling and puffing sounds which can be heard a great distance away in the heather or scattered woods, on the moorland and marsh which they inhabit. Even the slightest noise, however, can scare the birds away, leaving only a few feathers in the March snow.

← Capercaillie *(Tetrao urogallus)* at the display ground. Wingspan 5 ft; the cock weighs up to 11 lbs, the hen at its heaviest 6½ lbs. The Capercaillie is resident in the Pyrenees, Central Europe and Scandinavia and has been re-introduced to Scotland. In Scandinavia there are frequent hybrid birds, from the crossing of Capercaillie and Blackcock.

→ The display ceremonial of the Black Grouse *(Lyrurus tetrix)*, on ground covered with snow. The male is known as the Blackcock. Wingspan 3–6 ins; male weighs up to 4½ ins and the russet coloured female or Greyhen barely 1½ lbs. The Black Grouse is resident throughout Central Europe, England and Scotland and Wales, except for the South East, and in Scandinavia, except for the extreme North.

The Sousliks *(Citellus citellus)* live both above and under the ground. They are found mainly in the Balkans and the Ukraine, but recently they have spread to the Danube basin and to Czechoslovakia.

In Eastern Europe there are many rodents related to the squirrels, but which dig and therefore have less curved claws than the squirrels themselves. One distinct group of these rodents comprises the various species of Souslik. The Sousliks in fact belong to a group which includes the Marmots and Chipmunks, and although they are closely related to the squirrels, they do not live in treetops but dig long burrows in fallow land, pastureland and even at the edge of airports. These tunnels wind along for up to 24 ft and to a depth of 7 ft, and their digging operations turn over the soil thoroughly. Most colonies of Sousliks are found between the Volga and the Urals where live the Spotted Sousliks, which are somewhat smaller than the common Souslik. These do heavy damage in plantations of maize and are hunted in the same way as mice. Sousliks hibernate from November to April near the surface and collect large stores of food for the winter — but the yellow and black striped Souslik of Northern Russia, with its enormous cheek pouches, lays in the largest store of all these representatives of the Ground Squirrels, collecting up to 20 lbs of nuts, maize and beechmast in late autumn. While the Sousliks live mainly below the surface, the Flying Squirrels live above it and in the trees. The only European Flying Squirrel is found in mixed forests in the North of Siberia, Russia and Finland, living in holes in trees. This little 11 inch long animal with its dark brown "parachute" can glide through the air for up to 100 ft. It feeds nocturnally on fruits, berries, bark and insects.

100 years ago the Alpine Marmots *(Marmota marmota)* were only to be found in the Western Alps and in the High Tatra. Now they have been resettled in the Eastern Alps, the Carpathians and elsewhere.

The Bobac Marmot is found on the steppes of Southern Russia, but in smaller numbers than previously, and it has vanished from the steppes of Eastern Poland. The only other marmot in Europe is the Alpine Marmot, which is 24½ ins long, a fat animal found here and there in Central Europe, having been recently introduced into the Pyrenees and the Black Forest. The Bobacs decreased in number because of their valuable fur, but the Alpine Marmots were the victims of a superstition which held that because marmots withstand extreme cold high up in the mountains every winter, ointment made from marmot fat could cure rheumatism. However, this resistance to cold is due partly to the layer of fat and partly to the deep burrow with its thick lining of hay (up to 90 lbs). Of greater assistance in keeping warm is the complete change in the metabolism: the reduced activity of the thyroid gland produces a deep lethargy, slows down the circulation of the blood and reduces the heart beat to only three or four times a minute. During the six winter months the lungs contract and only take in as many breaths as they would in two summer days, for the sleeping marmot requires for this reduced activity but a small fraction of its normal amount of oxygen. During hibernation barely 15 grammes are burned up daily from the reserves of fat, but every few weeks a full bladder or a temperature below 39° forces the marmot into a short burst of activity to warm itself up. Within a few hours the body temperature rises by about 36°, but this warming up process uses up much more of the reserves of energy.

In hunting language the Roe Deer *(Capreolus capreolus)* is described as follows: the mothers *Does*, the young male and female *Kids* and the males *Bucks*. The yearling females have no special name in English.

About the first week in June, when the kids have been driven away by the doe, she retires into dense cover to drop the new fawns, often one of each sex. The rut has taken place during the previous June and July, but during the winter there is a pause in the development of the embryo which lengthens the gestation period by 3 months. Actually a second rut takes place in October but it is infertile. A new-born fawn only weighs 2 lbs, and 2 hours after its birth it can already stand upright on its thin legs, immediately making for the mother's nipples. As yet it cannot follow its mother on foraging expeditions, so it sleeps, hidden in the tall grass, while she looks for pasture. Its dappled coat blends with the sunlight and swaying flowers, so that only rarely does a bird of prey catch sight of it; also, the scent is so weak in the first few days that even fox and marten cannot smell it... the greatest danger to the fawn is from harvesters. When the fawn wakes and starts to search for its mother, glands between the toes of its hind legs secrete a delicate odour which the mother can follow through the grass. From the 8th day the fawn accompanies the doe about the territory. If it is a young male it soon produces very short antlers, called buttons, but only in the next year does the young buck grow antlers with a single point. Later each antler may have three points; this is the maximum. After the autumn rut the antlers are shed and grow again the following spring. The velvet is frayed off in April.

A group of piglets stare curiously into the camera. These are Wild Pig *(Sus scrofa)*, also known as Wild Swine. The mother is called a sow. The nose is known as the snout.

By the end of March the piglets, which were born 14 days earlier, venture out of their warm bed of leaves in the undergrowth of forests in Central and Southern Europe. When their striped coat is grown sufficiently to act as a camouflage among the brush, the piglets follow their enormous mother, who may weigh from 330 to 440 lbs, on her rooting expeditions. If one of the 6 to 12 piglets is threatened the sow will defend it madly and becomes even more dangerous than the tusked male boar. This huge male does not take part in the rearing of the young, but usually wanders about, restless and alert, over great distances. It frequently swims across rivers and lakes, and does not return to the sow until November to January, the time of the breeding season, when fights sometimes occur between young males and the older ones. The hide with its layer of fat has become as tough as armour-plating from the sap and resin of the boundary trees against which the boars rub themselves after their mudbaths, so that they can withstand even the most violent blows. Owing to the near-imperviousness of this hide to shot, the wild boar usually manages to escape the huntsmen, despite the lack of a close season for hunting. Farmers are continually trying to reduce the numbers of these pests which, as well as feeding on larvae and carrion, also eat a great deal of corn, turnips and potatoes. There are no wild boars in Britain or Northern Europe, as they were much hunted during the Middle Ages.

From 8 days old the young Elk calves frequently nibble at juicy birch leaves. For where wild Elks occur, see page 16.

Unlike the young of Roe and Red Deer, the Elk calves have no camouflage in the first few months of their life. They are like tiny imitations of their huge mother, the only female deer which is not smaller than the male. From its first pregnancy a cow will bear only one calf, but later there are usually two. The Elk cow is a solitary animal and its only companions are the young, whom it leads on foraging expeditions, clearing the way for the calves and making sure no wolves are approaching. But wolves are no match for the strong hooves of the Elk; only the Lynx and the Wolverine, which perch high up, have an advantage over it. Until the young can follow the mother everywhere, defence is better than flight. But a full grown Elk can escape at once by galloping confidently and swiftly over the soft marshland and powdery sand dunes, jumping over enormous holes and tree-trunks, even if these are six feet high. Nevertheless, when there are calves to be protected the cow remains with them to ward off danger. Elks are usually to be found in coniferous or mixed forests and particularly where there are stretches of water and marshland. They are close relatives of the American Moose, which is distinguished from the European Elk by its darker colour, more complex antlers and above all by its larger size.

The huge Elk cow *(Alces alces)* has a gestation period of 9 months and continues to suckle the calf into the winter.

The Elk cow sometimes treads down young leafy trees so that the calves can reach the fresh shoots at the top, but soon they must stoop to feed on plants. Adult Elks also need the juicy bark which they tear in long strips from the trunks of trees, to tan their hide. The young only begin feeding on bark, leaves and buds towards the end of winter when the cow's milk dries up because of the imminent birth of a new calf or calves. In spring the calves of the previous year leave their mother; after the birth of the new calves, many of the older ones try to attach themselves to their families once again, but they are seldom tolerated. The young males, who do not begin to grow horns until their second year, when they are just spikes 12 ins long, do not challenge Elks of higher rank in the hierarchy. The Elk's antlers do not begin to fork or branch until the third year, and later on they will be broad and palmate, with 3 to 12 points on each side: their size depends on the calcium content of the food and on the amount of sunshine to activate the vitamins which aid in bone formation. The adult Elk has long legs but a rather short body in comparison. Its muzzle is very long and hairy, and the Elk is also distinguished by the hairy dewlap on the throat. This is known as the "bell" because of its shape.

In past centuries bears often used to be hunted by parties of 300 men! In 1835 the last bear was shot in Bavaria; by 1882 there were no bears left in Austria; the last Swiss bear was killed in 1904 and since 1921 no more bears have been seen in the French Alps. Brown Bears became extinct in England in the 11th century. Only in Italy were any steps taken, by means of strict protective measures, to save the few surviving Alpine bears of Brento and Adamello in Western Trentino, and the remaining red-brown bears of the Abruzzi. Now, therefore, the shaggy-haired Brown Bears still move freely about their territory, 9½ to 12½ miles wide, like those in Scandinavia, Russia and the Balkans. Brown Bears, which may live up to 50 years, often vary considerably in the colour of the fur, stature and shape of the brow, and often even in behaviour. In Norway today they are reddish-brown with black feet; those in South Eastern Europe have pale ruffs, and in Southern Russia the coat is pale yellow. Normally the Brown Bear is a calm, cautious and phlegmatic animal, but if it is threatened or alarmed it immediately becomes a violent and aggressive opponent and is capable of moving with surprising speed and agility. Bears are also very good climbers.

7 year old Brown Bear *(Ursus arctos)* in Finland. A full grown bear, standing, may be 10½ ft tall and weigh 440 to 660 lbs. The Brown Bears of Scandinavia, Poland and Russia are nealy exterminated but those in Czechoslovakia and the Balkans are surviving very well.

The new born young of the Brown Bear are blind, rat-sized animals, weighing only 10 ozs. The female bears 2 to 3 in January, after a gestation period of 7 months. The rut can take place from April to June...experts are still not agreed about whether there is a pause in the development of the foetus, as in the Roe Deer, or delayed ovulation, as in the case of the Bats. The cubs remain blind for 4 to 5 weeks and stay with the mother in the warm winter lair for 10 weeks. It is not until their 4th month, when the female bear once again leaves the lair after a long period of fasting to feed on lichens and other plants, that the young ones follow her out. Although Brown Bears are carnivorous, the main food of the female and young consists of berries, fungi, roots and grasses as well as acorns, snails, beetle larvae and rodents. They are, of course, well known for their love of honey and frequently break into bee hives. They also sometimes plunder ant hills or catch fish in shallow mountain creeks. The adult Bear, however, may become truly carnivorous and has been known to kill cattle or sheep, Red Deer and Elks by breaking their back-bones. Sometimes they will start searching for food in the afternoon and continue right through until the next morning. During the winter, when they do not truly hibernate, Brown Bears are often seen outside their den, which is made of branches, lichen and moss.

Young Brown Bear with its white collar, which will disappear later, remaining only as a pale mark on the neck of the adult: resident Balkans and Carpathians. The young, whose tiny brain must increase in weight 38 to 58 times (four times as much as a man's), become sexually mature in their 5th year.

Both the Fallow Deer and the Mouflon are examples of animals whose range has been extended by man. The native home of the Fallow Deer is in Asia Minor, Northern Palestine and Southern Europe. However, the Normans introduced them into England and since the 11th century they have been flourishing in many European countries, where they added to the excitement of the hunt. Long before this they were dedicated to Artemis of Ephesus and were used as sacrifices by the Roman legions who spread them towards Gaul and Germania. As well as into Europe both bucks and does were introduced into North and South America, Australia and New Zealand. Today they are protected in all the countries where they are settled, for they enrich the landscape of the parks where they live. They feed mainly on grass, and the female bears one fawn in June or July after 8 months' gestation.

The Mouflon, however, is a native European animal which was rescued at the last moment from complete extermination: in the nature reserve of Cadarache in Corsica, on the Gennargentu Massive of Sardinia and on the island of Ponza in the Gulf of Gaeta, these wild sheep, which can leap great distances, still climb among the scrub and brushwood of the mountains. In 1869 Mouflon were introduced into the Danube region by Magyar princes, and today they are still found surviving even the hardest winters in Germany and Austria.

➤ Fallow Deer *(Dama dama)*, with its dappled summer coat, which gives way in late autumn to the blackish winter coat. From the 5th year onwards the antlers are narrow and palmed and not until the 10th to 12th year does the buck have any points on its antlers.

➤ Mouflon *(Ovis musimon)*. The ram's horns can weigh up to 11 lbs and are up to 32 ins long: ewes usually have either no horns at all or tiny ones only 2 to 2½ ins in length.

There are often 100 Chamois *(Rupicapra rupicapra)* in a herd. The family groups — mother, kid and yearling — remain close together.

There are seven races of Chamois in the mountains and in the upper limits of the tree belts of Europe. Their ancestors appeared at the end of the Tertiary period — 2 million years ago — and ranged from Asia Minor, over Thrace and Macedonia, into Europe. The descendants of these early Chamois pushed further northwards and westwards and were driven by the glaciers, which came and went 4 times in the 800 thousand years of the quaternary, to Belgium and Southern Italy. However, as the high mountains became green again, the Chamois returned, in particular to the Alps, and now reaches as far as Northern Yugoslavia. It has been successfully settled high up in the Black Forest, and has even been introduced to New Zealand. There are also the much smaller local race of fox-red Chamois in the Pyrenees, and the sandy yellow Cantabrian Chamois which only reaches a height of 28 ins at the shoulders. In the Abruzzi the Chamois has a dark grey collar on its pale winter coat. The largest Chamois live in the Carpathians, where an adult male may weigh up to 130 lbs. Its strongly curved horns can grow up to 14 ins long, while the weaker Caucasian variety have only short horns. The male Alpine Chamois has horns up to 9½ ins long, while the female's are thinner and shorter. The Chamois nearly died out in Switzerland at the end of the 19th century, but now there are 25,000 head in the country.

Every winter many Chamois are overwhelmed by snow. Weakened animals or the young of older parents rarely reach the next spring.

It seems incredible that Chamois manage to master their environment high up among steep rock walls and narrow ledges. They can leap easily over 20 ft ravines and can jump over 10 ft high boulders without a run, owing to the powerful leverage of their nearly perpendicular humerus and femur. Their steely but elastic hooves, cloven so that they can grip the smallest bump like prehensile pincers, enable them to find a firm foothold immediately after such leaps. Chamois can escape the showers of debris on the hillsides by their great speed: but if during their wild flight across the steep slopes one should slip, it can brake immediately by pressing its cloven hooves into the rock. In winter when the highest mountain ridges are frozen and slippery, the Chamois must move very carefully both on fresh glacier snow and that of the previous year; and where there are avalanches the herd must go in single file on its way down towards the mountain forests. Even if the temperature sinks as low as -22° C, the winter coat with its warm undercoat and long outer hairs keep the healthy Chamois from freezing. Normally they live on grass, leaves and fresh tree shoots: but in winter they may have to exist on twigs, bark and lichen scraped off rocks... and even then they usually reject hay which is dropped by helicopter or laid out for them, for it contains little nourishment. One animal always acts as a look-out when the herd is grazing — shrieking and stamping in warning.

Red Deer *(Cervus elaphus)* crossing a ford. Such large numbers are today only to be seen in Finland and Scotland.

The herds of stags and hinds have been diminishing in size throughout Central Europe, and only in the north and east can the large herds of graceful grey Atlantic Deer and strong Carpathian Deer still be found. When rutting takes place in autumn the herds scatter, and only reassemble after the stags have fought violently for the hinds. The victorious stags then stay with the hinds and the herds are led by an old hind. Only the stags have antlers, which they lose in February or March. The new antlers gradually start to develop again under the scalp which is rich in blood vessels. They grow for 80 days, ossify for another 40 days, cut off from the blood supply, and are fully developed by mid-August. The antlers are rubbed against bark until they become smooth and tanned. In the first year the antlers are simple and in the following years tines and then points are added. They reach their peak of development, usually not more than 18 or 20 points, at about 8 to 9 years old. Sometimes after the old antlers have fallen off a stag will gnaw at them, extracting their valuable calcium phosphate. The forming of new antlers depends on the functioning of the thyroid gland. There are many herds of Red Deer in England and Scotland, and they eat twigs, buds, leaves and grass, and even nibble bark off trees in winter. Red Deer measure from 73 to 84 inches long, the stag being considerably larger than the hind. In winter they have a greyish thick pelt which becomes smooth and brown in spring.

A herd of males of the Alpine Ibex *(Capra ibex)* fleeing over the frozen snow on rough, steely hooves.

Alpine Ibex do not flee blindly at the approach of men but wait until they can see and assess the danger. It is partly because of this confidence in their powers of getting away quickly and climbing great heights that they were shot in hundreds in the last century, until in 1861 they had virtually vanished from Switzerland, Austria and Germany. Only in Piemont did a tiny herd survive, a temptation to poachers who, after weeks of walking along mountain ledges, would return with their booty weighing 1½ cwt. Every part of the animal, from heart to horn, was regarded in folk medicine as a sure cure for almost everything, and therefore the hunters received a high price for their spoils. But then, in 1858, King Victor Emmanuel II of Italy made an end to the shooting, acquired the hunting rights in what is now the National Park of Gran Paradiso, installed 45 gamekeepers, and sent to prison for 5 years every poacher or *bracconiere* caught. 12 years later the herd had already grown to 400 head of Ibex and today it numbers over 4,000. From among these animals are descended all the present Alpine Ibex which have been reinstated among the fauna of Swiss mountain ranges (now 1,300) and of the Austro-Bavarian Alps (now 40). The same type of rescue operation was also carried out on the Pyrenean Ibex *(Capra pyrenaica)*, which has lyre-shaped horns and actually belongs to the Turs. These also had almost vanished from the Sierra de Gredos at the beginning of this century.

In the male herds of the Alpine Ibex two animals often engage in combat. During the day these big animals, weighing a minimum of 220 lbs, rest on rocks and ledges of cliffs. At sunset, however, they become active and set out to find pasture. Then fights may take place between younger and older males, both indulging in a sort of ritual display of strength, aggressiveness and stone-walling. One will suddenly rise up on its hind legs and dash its horns, which are covered with ridges, against the horns of its opponent. The other will stand firm and strike in its turn, making the sound echo among the mountain peaks. These fights are entirely harmless, and it seems as if the combatants are merely confirming their rank. For there is a long established hierarchy which is never disputed in such herds. However, when the rut takes place in December, the males are very calm, and there are no more fights... in any case, these would simply waste time, since the rut usually only lasts 4 days. For the photographer, one of the chief aims is to take a picture showing the almost legendary "*pas de deux*" which these animals frequently perform and which up to now has only been filmed once with Steinbocks in the Western Caucasus. Sometimes, if the need to escape arises, a male Ibex will spring from the edge of a perpendicular ravine and hurl itself downwards and across towards the opposite cliff face, thrust itself off again immediately on landing, and so zig-zag back and forth down the ravine to safety.

Europe's northern shore, the British Isles and Scandinavia, would be permanently ice-bound and populated with animals from the Arctic, if the temperature of the Gulf Stream were but 14° lower. The temperate climate, however, offers a suitable habitat for those animals which can avoid the yearly outbreak of winter, or are equipped to meet it. Those which remain active in snow and ice are naturally adapted to withstand the rigours of winter. It is only in those places where animals have been deprived of their normal winter sustenance by the cultivation of the soil and by forestry, that we must help and rescue them by supplying alternative sources of food. In the ancient forests of the nature reserves and national parks, however, where lush undergrowth penetrates the snow, and where bearded-moss rich in vitamins hangs from uncut branches within reach of Red Deer, Chamois and Ibex, no healthy animal need go hungry.

This photograph is more authentic and convincing evidence of the genuine struggle for existence, than one showing wild animals surviving thanks to provision of fodder by man. Here it may be seen that nature, when not interfered with, is perfectly capable of providing for wild animals and, by natural selection, keeps the different species strong and hardy. Yet this can only occur in completely primeval surroundings. Nothing that man provides can compensate the animals for the loss of their original habitat.

◄ At the end of the winter new dangers await those animals which have survived all the earlier hardships: often the wind-blown pasture is covered with soft new snow and every movement is difficult. The Red Deer of the Swiss mountains of Tamangur in the depths of Scarltal *(Grisons)* established their deer path in the Clemgia creek, which has so little water in winter that it hardly wets their legs. Suddenly, however, melting snow filled up the bed of the creek: but this did not deter the ten-point stag, which prefered to wade up to its breast in water in order to follow the current and to climb up the bank where the deep snow has already been trodden down by others. In a moment the Red Deer would have vanished round the bend of the little brook, which is overhung with snow. To take such photographs of animal behaviour one must have both patience and scientific knowledge!

NOTHING WILL BE LEFT / NOTHING UPON EARTH NOTHING UNDER THE EARTH / NOTHING IN THE WATERS ALL WILL BE HUNTED DOWN / ALL EXTERMINATED...

It was a close race, and Leonardo da Vinci's bitter prophecy could so easily have been fulfilled. For thousands of years Man struggled against a hostile Nature and defended his life against the overwhelming power of the beasts. That Man triumphed was to the credit of individuals; that superiority turned to arrogance was the guilt of the masses. It was the foresight of a few individuals who fought alone against the hordes of man to protect Nature and demanded laws to protect plants and animals against the manic destructiveness of the short-sighted. In 27 countries of Europe pioneer conservationists struggled for every inch of indigenous wild territory. Small are these sanctuaries—the ponds, the dunes, the cliffs—which nature-lovers with their small means were able to tear away from the profit hungry. But their example forced Princes and statesmen to heed the urgent warnings of the students of nature; and now understanding legislators have ensured that extensive nature reservations are protected. Long stretches of coastline, islands, moors, mountainous areas and virgin forests have become the last refuges of many animal species. For Man, too, these are oases of peace where he may wander and listen and learn again to watch. That the three following pages hardly suffice to name even the most important nature sanctuaries is welcome proof that the idea of nature conservation fell on fertile ground.

ALBANIA: DAJTI National Park, 7,300 acres, 15 miles east of Tirana.—DIVJAKA National Park, 4,800 acres, 17 miles west of Lushnje on the Adriatic Coast.—LURA National Park, 7,300 acres, between 3,000 and 7,000 feet above sea level, 25 miles north-east of Burreli: lynxes and bears.—TOMORI National Park, 7,300 acres, 2,500 to 8,000 feet above sea level, 15 miles east of Berat.

AUSTRIA: KARWENDEL Protection Zone, 175,000 acres, in Tyrol.—BÖHMERWALD, a National Park of 230,000 acres in Upper Austria: Eagle Owl eyries.—HINTERSTODER PRIEL, 145,000 acres, National Park in Upper Austria: chamois, eagles.—The high plateau of the STEINERNES MEER and HAGENGEBIRGE, a protected area of 85,000 acres: ibexes, chamois.—SEEWINKEL Nature Reserve and NEUSIEDLERSEE Park, 400 acres and 85,000 acres, in Burgenland on the Austro-Hungarian frontier: breeding colonies and winter quarters of water birds and hundreds of herons and spoonbills. There is a plan to turn the whole lakeland area into a National Park.—TAUERN Nature Reserve and Park, 9,750 acres, and the GROSSGLOCKNER-PASTERZER area are the centres of a planned National Park in Hohe Tauern.—ROTHWALD Nature Reserve, 1,450 acres, near Lunz-am-See in Lower Austria is the largest virgin forest in central Europe.—LOBAU, WACHAU and WIENERWALD are Nature Reserves: it is planned to turn the banks of the Danube into a National Park.—DONAU-AUEN, 3,300 acres between the two rivers Traun and Enns in Upper Austria: herons, cormorants, and kites as well as fallow deer. There are also a great many smaller reserves.

BELGIUM: LESSE ET LHOMME National Park, 2,400 acres, in Namur Province.—BIRD PROTECTION AREAS: The whole of the north-east Belgian coast is a bird sanctuary where shooting is forbidden.—ZWIN, ZOUTE, and HAZEGRAS, 4,150 acres, sea bird nesting places (96 different species) on the coast of Flanders.—DE PUTSMOER, 1,800 acres, in the province of Antwerp near Kalmthout: breeding ground of rare Shelduck.—DE ZEGGE, 80 acres, in Antwerp Province: protected area for marsh birds.—BERENDRACHT, a heron colony just north of Antwerp, breeding ground for 250 pairs of grey herons.—In addition there are 23 smaller reserves.

BULGARIA: VITOCHA National Park, 150,000 acres, south of Sofia.—KILIAKRA Reserve in South Dobrudscha; breeding grounds of Black Sea seals.—MILKA, a bird sanctuary near Svistov on the Danube.

CZECHOSLOVAKIA: KRKONOSE and JIZERSKÉ HORY National Parks—375,000 acres in all, of which 100,000 are an absolute sanctuary.—TATRA National Park, 120,000 acres of open parkland, 170,000 acres of protected zone, adjoining the Polish National Park of the same name: chamois, wild cats, bears, wolves.—KARLSTEJNSKO, 3,700 acres, and KODA, 1,100 acres, south-west of Prague: fauna of the Steppes.—BOZI DAR Reservation, 3,700 acres in the Erzgebirge.—Also 300 smaller nature reserves.

DENMARK: JORDSAND and VALDEHAVET, 25,000 acres, North Friesian Islands, are banned for shooting.—DYREHAVEN, 4,800 acres, near Copenhagen is a National Park.—VORSØ ISLAND, 140 acres, is a sea bird sanctuary in Horsensfjord, as well as HIRSEHOLMEN, a group of islands in the Kattegatt near Frederikshaven.—TIPPERNE and KLAEGEBANKEN, 2,150 acres, a group of islands in Ringkøbingfjord on the west coast: 50,000 wild ducks.—GRAESHOLMEN Island, east of Bornholm in the Baltic: large colony of Eider Ducks, Razorbills, and Guillemots.—In addition to these there are 158 smaller reserves.

EAST GERMANY: SAXONIAN National Park "Sächsische Schweiz" of 336,000 acres.—MÜRITZSEE Nature Reserve, 16,000 acres, near Mecklenburg, the home of Germany's largest crane colony.—Beaver Protection Areas: GREATER PINNOWSEE near Bernau and LÖDDERITZ-STECKBY near Schonbeek.—Bird Protection Areas: LANGENWERDER near Wismar, Mecklenburg Bay; CONVENTERSEE, east of Heiligendamm on the Baltic Sea coast; JASMUND on Rugen Island; DER DARS, a 9,600 acre peninsular south-west of Rugen. In all 210 reserves.

FINLAND: National Parks in Lappland: PALLASOUNASTUNTURI, 120,000 acres on the river Ounasjoki.—PISAVAARA, 13,000 acres on the river Kemi-joki.—MALLA with Lake Kilpisjärvi, 7,200 acres, on the Norwegian-Swedish frontier. All three are a refuge for sub-arctic fauna, bears, wolves, lynxes, wolverines, elks etc.—OULANKA National Park, 24,600 acres, a wild river area of Oulankajoki: elks, deer.—LEMMENJOKI National Park of 91,000 acres: beavers, minks, otters.—LINNANSAARI National Park, 2,000 acres: 20 islands in Lake Haukiivesi of the Finnish lake plateau: ospreys.

FRANCE: THE CAMARGUE: A botanic and zoological Reserve of 33,000 acres in the Rhone delta: flamingo colony, resting place and winter grounds of numberless sea birds and 100,000 wild ducks; wild bulls and herds of horses.—PELVOUX National Park, 67,000 acres, in Isère and Basses-Alpes Departments.—LAUZANIER Reserve, 7,200 acres, Basses-Alpes Department on the Italian border.—NÉOUVIEILLE Reserve, 5,250 acres, in the Pyrenees: 25 lakes; vultures.—8 Mile long Beaver Protection Area at the mouth of the river SÈZE.—FONTAINEBLEAU FOREST including 9,650 acres of complete sanctuary and 2,350 acres of park lands.—Bird Protection Areas: In all there are 1,500, of which the most important is: SEPT-ILES, ornithological reserve comprising 55 acres of land and 2½ square miles of sea off the north coast of Brittany: breeding colonies of many sea birds and puffins. *Corsica:* Cadarache Reserve, the traditional home of the wild sheep (Moufflon). *In creation:* CAROUX National Park, Hérault Department in the South of France.—BORÉON Reserve in the Alpes Maritimes: chamois and ibexes.—VANOISE National Park in Haute Savoie, a French extension of the Italian Gran Paradiso National Park.

GREAT BRITAIN: *Scotland:* CAIRNGORMS National Reserve of 39,689 acres: introduced reindeer, eagles, ospreys, deer, wild cats.—ISLAND OF RHUM, 26,400 acres, an island sanctuary south of the Isle of Skye: breeding ground of seals, 1,500 red deer and a colony of puffins.—BEN EIGHE, 10,450 acres, National Reserve in Rosshire: wild cats, deer, and eagles.—HERMANESS ISLAND in the North Shetlands, 2,383 acres: sea birds and seals.—SAINT KILDA'S ISLAND National Reserve, 2,107 acres, breeding colony of fulmars, puffins, 7,000 pairs of kittiwakes, 13,000 pairs of guillemots, 40,000 pairs of gannets.—*England:* (North to South). FARNE ISLANDS, 80 acres off the north coast of Cumberland and THE CALF OF MAN, 620 acres, off the Isle of Man: sea bird colonies.—THE LAKE DISTRICT, a National Park of 564,240 acres, comprising parts of Lancashire, Cumberland and Westmoreland.—SNOWDONIA, 560,800 acres, in North Wales.—THE PEAK DISTRICT, 346,880 acres, in Derbyshire.—THE NORTH YORKSHIRE MOORS National Park of 353, 920 acres.—OXFORDNESS and HAVERGATE Nature Reserve, 514 acres, an island off the Suffolk coast with 154 different species of birds.—PEMBROKESHIRE COAST National Park, 144,000 acres, in North Wales.—SKOMER ISLAND off the PEMBROKESHIRE coast, 722 acres: puffin colony and a breeding ground for grey seals.—BRIDGEWATER BAY, 6,076 acres, in the Bristol Channell: herons and wild geese.—DARTMOOR National Park, 233,600 acres in Devonshire. *N. B. also: THE WILDFOWL TRUST at Slimbridge on the Severn Estuary in Gloucestershire and at Peakirk in Northants, two of Great Britain's most important wild fowl gardens.*

GREECE: PARNASSUS National Park, 14,500 acres, in North Attica.—OLYMPUS National Park, 9,600 acres, between Macedonia and Thessaly.—KOTZA-ORMAN virgin forest reservation of 4,500 acres in Macedonia.—The Island of Ghyaros, 2,160 acres, in the Northern Sporades. ATALANTONISI Island near Atalanti. THEODOROU Island of 200 acres north of Crete. ANTIMILOS Island. All four are proteceted areas for rare Greek wild goats.—SAMARIAS, 2,000 acres, in Crete: rare Cretan chamois.

HUNGARY: KISBALATON, 6250 acres, a bird sanctuary on Lake Balaton. Breeding colonies of herons and ibises, winter quarters for water birds.—TIHANY National Park, 1,700 acres, in the Balaton district. FÉHÉR Reserve (On the White Lake), 620 acres, on the Jugoslav-Hungarian border: salt marshes, water birds.

ICELAND: THINGVELLIR National Park, 7,450 acres, 30 miles east of Reykjavik: arctic foxes, snowy owls, Greenland falcons. Also 5 other Reserves, totalling 37,000 acres.

IRELAND: NORTH BULL ISLAND, 3,500 acres, on the east coast near Dublin: sea birds.—LOUGH STRANGFORD, lagoon-sanctuary with 17 islands on the east coast of County Down: sea birds, seal breeding grounds.—THE SKELLIGS, bird sanctuary on the south-west coast of County Kerry: breeding colony of 10,000 pairs of gannets.—LAMBAY ISLAND, 620 acres, on the east coast north of Dublin: one of the grey seal's most southerly breeding grounds.—Also, AVONDALE, 260,000 acres, in County Wicklow, and BORNE VINCENT MEMORIAL PARK, 11,000 acres, near Killarney.

ITALY: STELVIO National Park, 240,000 acres, Trentino, Alto-Adige (South Tyrol): red deer, chamois, Alpine Brown Bears.—GRAN PARADISO National Park, 140,000 acres, in Piemont: 4,000 ibexes.—ABRUZZI National Park, 73,000 acres, near Pescasseroli: chamois, wolves, Abruzzi-bears, wild cats.—CIRCEO National Park, 1,850 acres, coastal sanctuary near Sabaudia, south-east of Rome: South European wild boars and porcupines.—LA MANDRIA, 6,500 acres, near Turin: reservation for stags and migratory birds.

JUGOSLAVIA: MAVROVO National Park, 175,000 acres, in Macedonia.—TARA National Park, 73,000 acres, in

Serbia.—PLITVICKA JEZERA National Park (The Plittwitzer Lakes), 36,000 acres, in Croatia.—PERISTER National Park, 25,500 acres, near Lake Prespa, Macedonia.—PAKLENICA and RISNJAK National Parks, 6,750 acres, on the Adriatic coast border of Croatia.

LUXEMBURG: HAUTE-SURE National Park (planned).

NETHERLANDS: HOGE VELUWE and VELUWEZOOM National Parks, 14,500 acres and 9,750 acres, in east Gelderland: red deer and wild sheep.—GOOIS Nature Reserve, 4,500 acres, between Huitzen and Hilversum. VENNEN VAN OISTERWIJK, 3,600 acres, and STRABRECHT, 3,000 acres, in North Brabant.—KENNEMER DUINEN, 3,000 acres, near Haarlem. Bird Protection Areas: many islands totalling 24,750 acres, and especially: BOOSPLAAT, 11,500 acres, and NOORDSVAARDER, 1,250 acres, on Terschelling Island. GEUL EN WESTERDUINEN, 3,600 acres, and MUY EN SLUFTER, 1,900 acres, both on Texel Island. MEEUWENDUINEN, 900 acres, on Vlieland Island.—Also DE BEER, 2,150 acres, off the Hook of Holland.—NAARDERMEER, 1,800 acres, south-east of Amsterdam, marshbird sanctuary: cormorants, herons, etc. HET ZWARTE MEER, 5,000 acres, in the southern part of North East Polders.—On the Zuidersee Polders the reclaimed land is being observed according to Darwin's theories. The Netherlands possess in all 200 smaller reserves comprising some 150,000 acres.

NORWAY: NORDMARKA Reserve, 7,000 acres, near Oslo. —FOKSTUMYRA Reserve, 2,300 acres, in Opland: wild reindeer, deer and elks.—VAGGETEM Reserve, 1,200 acres, on the Russo-Norwegian border: sub-arctic fauna.—Sea-bird Reserves: TOFTEHOLMEN, MOLEN ISLAND and RANVIKHOLMENE, long stretches of coast-line and groups of islands in Oslo-Fjord.—In Norway Nature Protection has only just begun, as immense stretches of country are still in a virgin state.

POLAND: KAMPINOS National Park, 98,000 acres, of which 45,000 are an absolute sanctuary, near Warsaw: elks, herons, cranes.—HOHE TATRA National Park, 50,000 acres, south of Cracow on the Polish-Czech border: ibexes, bears, wolves, lynxes.—BIALOWIEZA National Park, 25,000 acres, south-east of Bialystok on the Polish-Russian border: virgin forest and marshland, bison, elks.—PIENINACH National Park, 5,000 acres, north-east of Hohen Tatra on the Czech border.—BABIAGORA National Park, 4,000 acres, in the Carpathians: ibexes and chamois. There are a further 450 smaller reserves and two Nature Protection Areas are planned: WOLIN, a lagoon on the Baltic coast near Swinoujscie (Swinemünde), shifting sand dunes, breeding colonies of sea-birds; LEBA, an inland sea, and GARDNO, 80,000 acres, on the Baltic north-east of Stupsk (Stolp): cormorant colonies.

PORTUGAL: GERES National Park, between the rivers Cavado and Lima on the northern Spanish-Portuguese border.

RUMANIA: RETEZATU National Park, 24,000 acres, West Transylvanian Alps: wolves, lynxes.—PIETROSUL MARS DELA BORSA National Park, in Carpathia—no boundaries: chamois, Carpathian deer, wolves.—DANUBE DELTA and BRAILASMARSH, Bird Protection Zone: pelicans, herons, ibises.

SPAIN: SAJA Y AGREGADOS Reserve, 145,000 acres, between 650 and 6,500 feet above sea level, south-west of Santander.—ANAYET Reserve, in the Pyrenees: vultures, wild boar, chamois, ptarmigan.—CAVADONGA National Park, in Oviedo Province, west of Picos d'Europa in Northern Spain.—RERES Y BRANAGALLONES, a National Park in the Cantabrian mountains: 1,000 chamois.—GREDOS National Park, in the Province of Avila: Spanish ibexes.—SERRANIA DE RONDA, 50,000 acres, National Park, Cordilleres, Malaga Province: wild goats and a small southern European species of deer.—TEIDE National Park, 27,000 acres, in the Canary Islands: a mountain of over 12,000 feet and the crater of the Teide Volcano.

SWEDEN: 6 large National Parks in Lappland, rich in subarctic fauna and also bears, wolves, lynxes, wolverines: SAREK, 475,000 acres; STORA SJÖFALLET, 415,000 acres; MUDDUS, 120,000 acres; PELJEKAISE, 35,000 acres; ABISKO, 12,000 acres; VADDETJAKO, 6,500 acres. 71 bird sanctuaries covering 1,250,000 acres, inland seas, long stretches of coast line and small archipelagoes in the Baltic, in Kattegatt and Skagerak, of which the largest, SJAUNJA, alone covers 720,000 acres.

SWITZERLAND: THE SWISS NATIONAL PARK, 42,500 acres, in Lower Engadine on the Italian Border: 150 ibexes, chamois, mountain deer, and a rich variety of Alpine flora and fauna.—ALETSCH Reserve, 75,000 acres, in Valais. —GRIMSEL Reserve, 4,800 acres, in the Berneses Oberland. —35 confederate and 75 cantonal nature reserves exist, in addition to more than 20 smaller nature reserves and 30 bird sanctuaries.

U.S.S.R.: DARVINSKI Reserve (named after Darwin!), 430,000 acres, on the Upper Reaches of the Volga; protected area around the basin of the Rybinsk barrage, used for studying the adaptation of flora and fauna to changes in environment.—*White Russia:* BELOWEZHSKAYA PUSHCHA Wildlife Park, 180,000 acres, adjoining the Polish National Park: bison, wolves, bears, elks, red deer.—BERESINA National Park, 160,000 acres, Beresina River district: beavers, otters and elks.—*Kola Peninsula:* KANDALAKSCHA Reserve, 50,000 acres, stretches over 32 islands of the White Sea and over 6 islands of the Barent Sea east of Murmansk, breeding grounds of sea-birds and seals.—*Lithuania:* ZHUVINTAS Reserve, 7,400 acres, north of Kaliningrad (Königsberg): breeding colonies of wild mute swans.—*Don:* VORONEZH Reserve, 95,000 acres; beavers and Musc shrews in a wild state, elks; Large-scope breeding of Old World Mink and Sable.—*Ukraine:* KRYMSKI Reserve, 73,000 acres, south-east of the Krim Peninsular: red deer and wild sheep (moufflons).—TCHORNOMORSKY, 29,000 acres, long coast line and islands in the Dnjeper Delta: sea and marsh birds, herons, ibises and pelicans.—*Caucasus:* KARKAZKI National Park, 245,000 acres: bison, chamois and bears.—*Volga Delta:* ASTRACHAN Reserve, 100,000 acres, bird sanctuary on the Caspian Sea.—*Aserbeidschan:* KYZYL-AGACHSKI Reserve, 200,000 acres, in the Kura Delta, south of Baku, on the Caspian Sea. The third flamingo colony on the edge of Europe.

WEST-GERMANY: *Baden-Würtemberg:* HOLZMADEN, 19,000 acres, where fossil flora and fauna of the Jurassic era, 165 million years ago, are preserved.—FELDBERG, 8,000 acres, a nature reserve in the Black Forest.—*Bavaria:* KARWENDEL and KARWENDELVORGEBIRGE, 65,000 acres, a nature reserve near Garmisch-Partenkirchen.—KÖNIGSEE AREA, 49,000 acres, near Berchtesgaden, the centre of the Austro-German planned National Park Partnership in the Steinernes Meer and the Königsee.—AMMERGEBIRGE, 48,000 acres, nature reserve.—*Hesse:* HOHER VOGELSBERG, 52,000 acres, a natural park north of Frankfurt am Main.—KÜHKOPF-KNOBLAUCHSAUE Nature Reserve, 5,700 acres, near Gross-Gerau: water birds and a heron colony.—*Lower Saxony:* LÜNEBURG HEATH Nature Park, 48,000 acres.—MÜNDEN Nature Park, 47,000 acres, including 125 acres of moorland. DIE LUCIE, 4,350 acres, near Dannenberg: swamp forest with water birds.—*North Rhineland and Westphalia:* SIEBENGEBIRGE Nature Reserve, 10,000 acres, near Bonn.—KRIECKENBECKER LAKES, 2,000 acres, the winter grounds of water birds.—*Rhineland and Palatinate:* PFÄLZER WALD Nature Park, 400,800 acres.—SÜDEIFEL Nature Park, 25,500 acres, near Bitburg: red deer and fallow deer.—*Schleswig-Holstein:* HAHNHEIDE, 3,500 acres, near Trittau.—Bird Protection areas: LISTER DUNES and ELLENBOGEN PENINSULA, 4,850 acres, on Sylt Island.—There are 690 smaller nature reserves in West Germany.

For their help in the compilation of this survey we thank the International Union for the Conservation of Nature and its Resources, Morges, Switzerland, and the London office of the World Wildlife Fund, 2 Caxton Street, London S. W. 1

(Acreages are approximate)

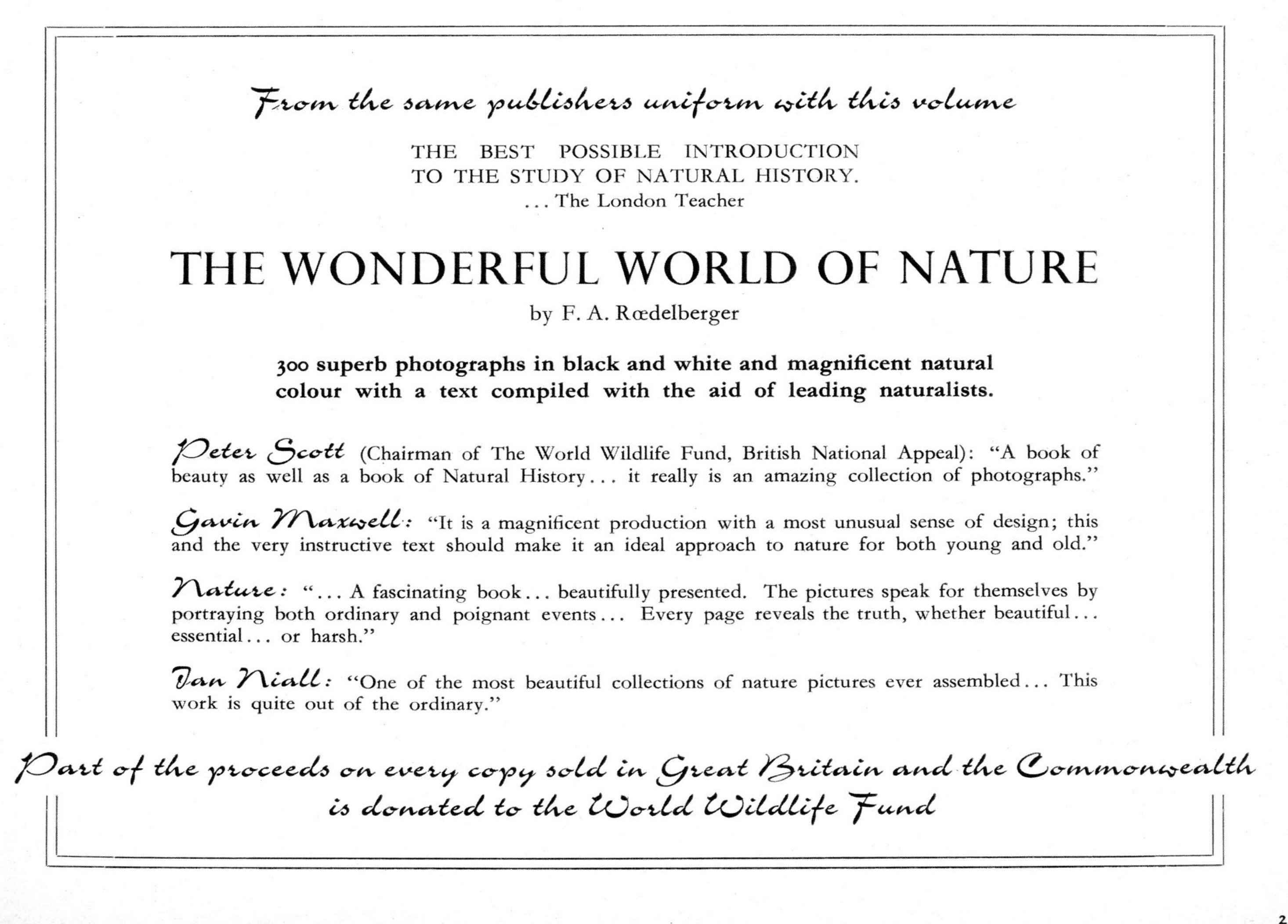

This English version is based on the Swiss original bilingual edition:

"FÉERIE ANIMALE" / "TIERWELT EUROPAS"

ALPHABETICAL INDEX OF ILLUSTRATIONS

INDEX OF CONTRIBUTORS

The Editors and the publishers thank the Natural History Museum, Berne and Dres. Walter Huber and Hannes Sägesser, for their great help in the publication of this book.

The publishers are particularly grateful to Dr. Bruce Campbell, Dr. H. Gwynne Vevers, E. A. Ellis, and John Markham, F.R.P.S., F.Z.S., for their assistance in the preparation of the English language edition.

They acknowledge a debt to the vast accumulation of scientific observation and knowledge built up by many naturalists down the ages, and especially to these works:

Prof. Dr. Paul Bröhmer "Die Lebensgemeinschaften der Gewässer", Tony Burnand und Godefroy Schmid "Le Grand Livre de la Mer et des Poissons", Paul Géroudet "Les Rapaces", Prof. H. H. Gigliolo: Proceedings of the Scientific Meetings of the Zoological Society of London 1889, H. Gouttière "Le Monde vivant", C. A. W. Guggisberg "Das Tierleben der Alpen", Werner Haller "Geheimnisvolles Federvolk", Frank W. Lane "Kingdom of the Octopus", Prof. Dr. Günter Niethammer "Tierausbreitung", Peterson-Mountford-Hollom "A Field Guide to the Birds of Europe", Prof. Dr. Adolf Portmann "Meerestiere und ihre Geheimnisse", "Das Tier als soziales Wesen", "Von Vögeln und Insekten", Paul A. Robert "Les Insectes", Hans-Wilhelm Smolik "Das grosse illustrierte Tierbuch", Touring Club Italiano "La Fauna".

Over 100 internationally famous wildlife photographers have contributed to this collection:

BLACK AND WHITE PHOTOGRAPHS

COLOUR PHOTOGRAPHS

ROTOGRAVURE AND COLOUR-HELIOGRAVURE: VERBANDSDRUCKEREI LTD BERNE

Printed in Switzerland